A C S S Y M P O S I U M S E R I E S **479**

Element-Specific Chromatographic Detection by Atomic Emission Spectroscopy

Peter C. Uden, EDITOR

University of Massachusetts at Amherst

Developed from a symposium sponsored
by the Division of Analytical Chemistry
at the 199th National Meeting
of the American Chemical Society,
Boston, Massachusetts,
April 22–27, 1990

American Chemical Society, Washington, DC 1992

Library of Congress Cataloging-in-Publication Data

Element-specific chromatographic detection by atomic emission
  spectroscopy /   Peter C. Uden, editor

  p.   cm.—(ACS Symposium Series, ISSN 0097–6156; 479).

  "Developed from a symposium sponsored by the Division of
Analytical Chemistry at the 199th National Meeting of the American
Chemical Society, Boston, Massachusetts, April 22–27, 1990."

  Includes bibliographical references and index.

  ISBN 0–8412–2174–X

  1. Chromatographic   analysis—Congresses.   2. Atomic   emission
spectroscopy—Congresses.

  I. Uden, Peter C., 1939–   II. American   Chemical   Society.
Division of Analytical Chemistry.   III. American   Chemical   Society.
Meeting (199th: 1990: Boston, Mass.) IV. Series.

QD79.C4E44   1991
543'.089—dc20   91–37718
                                          CIP

The paper used in this publication meets the minimum requirements of American National
Standard for Information Sciences—Permanence of Paper for Printed Library Materials, ANSI
Z39.48–1984.   ∞

Copyright © 1992

American Chemical Society

ACS Symposium Series

M. Joan Comstock, *Series Editor*

1992 ACS Books Advisory Board

Foreword

THE ACS SYMPOSIUM SERIES was founded in 1974 to provide a medium for publishing symposia quickly in book form. The format of the Series parallels that of the continuing ADVANCES IN CHEMISTRY SERIES except that, in order to save time, the papers are not typeset, but are reproduced as they are submitted by the authors in camera-ready form. Papers are reviewed under the supervision of the editors with the assistance of the Advisory Board and are selected to maintain the integrity of the symposia. Both reviews and reports of research are acceptable, because symposia may embrace both types of presentation. However, verbatim reproductions of previously published papers are not accepted.

Contents

Preface

CHROMATOGRAPHERS AND OTHER SEPARATION SCIENTISTS in the diverse fields of chemical and biological sciences face two primary problems as they consider procedures to accomplish their analytical goals. The issue of resolution of the sample components is always a central concern; however, the identification and overall characterization of separated eluates have also become vital tasks. Detection systems were first conceived purely as a means to monitor an eluent stream for component content and level, but those detectors that responded to specific sample properties rather than merely to a change in a bulk physical property of the mobile phase were found to afford a valuable means of characterizing sample eluates.

Although detection systems were first devised as integral units within the chromatograph, an increasing proportion of detection now involves combined, tandem, or interfaced "characterization detectors," which are analytical measuring instruments optimized for on-line measurements of flowing eluent streams. These powerful combinations call for sophisticated designs because the chromatograph, the instrumental detector, and the chromatograph–detector interface must all be optimized and integrated.

The most extensive developments in interfaced detectors have been in gas chromatography (GC), in which the absence of a condensed mobile phase often permits efficient analyte transfer to the detector while minimizing detection interferences caused by the mobile-phase background. Even so, many effective high-performance liquid chromatographic (HPLC) and supercritical-fluid chromatographic (SFC) interfaces have been devised. The most powerful interfaced-detection techniques are the various analytical spectrometries, each of which discloses specific characterization information on the eluates. The most widely used interfaced-detection technique is mass spectrometry (MS), which reveals molecular information. In GC, bench-scale instruments, in which the spectrometer is a truly integrated part of the chromatograph, are common. Fourier-transform infrared instrumentation has likewise facilitated the development of GC–IR spectroscopy for functionality characterization. MS and, to a lesser extent, IR detectors are also used extensively in HPLC and are feasible for SFC.

Elemental analysis is the particular strength of atomic spectroscopy. The renaissance of analytical atomic emission spectroscopy (AES) during the past decade, a result of the practical development of analytical plasmas such as the inductively coupled plasma (ICP) and the microwave-induced plasma (MIP), has made element-specific chromatographic detection feasible for all elements. Although atomic spectroscopy is considered more the province of the inorganic analyst, and chromatography (particularly GC) is considered that of the organic analyst the interfacing of chromatography with atomic emission spectroscopy (C–AES) has wide applicability for various analytical characterizations.

The C–AES field has grown extensively, as evidenced by the number of papers and reviews being published and symposia being presented. An important development has been the introduction of commercial instrumental systems for GC–AES; these systems make the C–AES technique available to a wider audience. Parallel developments for HPLC are close, and the wider adoption of sophisticated techniques such as ICP–MS has enabled the successful exploration of such techniques as HPLC–ICP–MS.

The state of the science of C–AES and its versatility were covered in depth at the symposium upon which this book is based. Scientists from academic, industrial, and instrument company laboratories shared both their previous achievements and their expected goals.

The goal of this volume is a practical one. The field of C–AES has much to offer to the practicing analytical chromatographer and atomic spectroscopist. My aim was to assemble information that is both fundamental as source material and of value in analytical problem solving. I hope that this is not just a symposium collection but a compendium of useful applications.

I thank the authors for their contributions, either for participating in the symposium or for being willing to enhance this volume by contributing an invited chapter. I also acknowledge the Donors of the Petroleum Research Fund, administered by the American Chemical Society, for partial support of the symposium and for enabling overseas participants to attend by providing financial assistance.

PETER C. UDEN
Department of Chemistry
University of Massachusetts
Amherst, MA 01003

August 21, 1991

Shortly after the symposium was held, we were all saddened to learn of the death of Professor Peter N. Keliher, one of the symposium participants and a chapter contributor. Peter Keliher was an atomic spectroscopist who was one of the first to recognize the strength of interfaced C–AES and who made important contributions to that field as well as more widely to the discipline of analytical chemistry. All too soon we lost a fine scientist, teacher, and scholar, and a good friend.

Chapter 1

Atomic Spectral Chromatographic Detection
An Overview

Peter C. Uden

Department of Chemistry, University of Massachusetts, Amherst, MA 01003

The development and current status of atomic emission spectroscopy (AES) as utilized in element selective detection systems for chromatography is reviewed. The problems of interfacing the eluent streams for gas chromatography (GC), supercritical fluid chromatography (SFC), and the various modes of high performance liquid chromatography (HPLC) are indicated and the various approaches to their solution are evaluated. The analytical performance of different chromatography - atomic emission spectroscopic (C-AES) detection systems is compared and the sensitivity, selectivity and dynamic ranges of instrumentation for elements that have been investigated are reviewed. Selected applications of historical, developmental and current interest are included.

The concepts and implementation of chromatographic detection have always stemmed from the notion of continual qualitative and quantitative monitoring of flowing stream composition. All chromatographs involve a detection device to measure the components resolved by the column; this detector should respond immediately and predictably to the presence of solute, at low concentrations, in the eluting mobile phase. One class of detectors, 'the bulk property detectors' respond to changes produced by eluates in a characteristic mobile phase (gas, liquid or supercritical fluid) physical property. These detectors are effectively 'universal' but they are typically relatively insensitive, since their response is characterized by small relative changes induced by low concentrations of analytes present in the bulk flowing eluent stream; further, they give little direct information about the separated chemical species. A second group of detectors, the 'solute property detectors', respond directly to some physico-chemical property of the eluates. The latter detectors, which can give

0097–6156/92/0479–0001$07.00/0

'selective', or 'specific'. information on the nature of eluates, play an important role in the modern laboratory.

Analytical spectroscopy has assumed an increasingly important place in chromatographic detection, because of its capacity to present an 'information-rich' spectrum of separated compounds, as well as acting only as a monitoring device. Spectral property detectors such as the mass spectrometer, the infrared spectrophotometer and the atomic emission spectrometer fall in this class. These detectors are 'property selective', 'structure or functionality selective' or 'element selective', and give the analyst eluate characterization based on these physico-chemical features. Measurement of properties, structure, functions and elemental composition directly within the chromatographic separation process broadens the analytical role of chromatography considerably.

The objective of element selective chromatographic detection is to obtain qualitative and quantitative information on eluates, frequently in the presence of interfering background matrixes, by virtue of their elemental constitution. Further, by analogy with classical 'elemental microanalysis', simultaneous multielement detection can enable empirical formulae of eluates to be determined. Element selective detectors used in GC include the alkali flame thermionic detector (AFID or NPD) for nitrogen and phosphorus, the flame photometric detector (FPD) for sulfur and phosphorus and the Hall detector for halogens, nitrogen and sulfur. However they are only usable for those particular elements, and the concept of a device which could detect any desired element specifically (the periodic table detector, the PDT) is attractive.

Atomic spectroscopy may be recognised as the most fundamental analytical technique for elemental determination, but while the paths of chromatographers and of atomic spectroscopists sometimes cross, these groups of physical scientists have less frequent contact than might be wished from the viewpoint of analytical chemistry. This is not surprising because inorganic determination is the major goal of analytical atomic spectroscopy, while most chromatographic methods focus on organic, biochemical or polymer separations and analysis. However, consideration of the nature of detection devices used for analytical chemical separations, suggests that present-day instrumentation relies heavily of the conceptual integration and physical interfacing of, what may be at first sight, quite dissimilar and independent techniques. Thus the time was ripe for the instrumental concept of C-AES (Chromatography - Atomic Emission Spectroscopy) as the need for wider elemental chromatographic detection arose.

It is worth repeating the observation on the different perspectives of an instrument which interfaces the chromatograph with a complex 'sample characterization device' such as the atomic emission spectrometer. To the chromatographer, the spectrometer is a sophisticated 'chromatographic detector', but from the spectroscopists point of view, the chromatograph is a component-resolving sample introduction device. Both observations are true and emphasize that this mode of chemical analysis must consider and optimize both the separation and the detection process, as well as the 'interface' or link between them.

Different types of atomic spectroscopy have been interfaced for chromatographic detection. Among these are atomic absorption (AAS), flame emission (FES), atomic fluorescence (AFS), and atomic emission (AES) spectroscopies(*1,2*). The capability of AES for simultaneous multielement measurement, while maintaining a wide dynamic measurement range and good sensitivities and selectivities over background elements, has led to atomic plasmas becoming very widely used during the 1980s. It is this technique which has been been most studied for chromatographic interfacing.

The advantages of C-AES with plasmas may be summarized as follows: i) monitoring the elemental composition of eluates directly with high elemental sensitivity; ii) monitoring for particular molecular functionality by means of specific reactions with derivatives containing element 'tags'; iii) toleration of non-ideal chromatography, the inherent selectivity of plasma emission enabling incomplete chromatographic resolution from complex matrixes to be overcome; iv) simultaneous multi-element detection for empirical and molecular formula determination.

Quantitative detection requires full sample transfer from column to plasma, although an ancillary detector may be placed in series (if it is non-destructive) or in parallel, through appropriate effluent splitting, with the atomic plasma. The high temperature of the plasma (between 2000°C and 8000+°C, depending upon the plasma type, provides enough energy to break down all molecular species to their constituent atoms and derived ions, which are excited by the plasma energy and subsequently emit radiation at spectral wavelengths characteristic of the particular element. The essence of the plasma spectral detector is that elemental response should be independent of molecular configuration; this is true in most situations although in some cases, for specific plasmas, molecular or chemical reactivity dependence of response has been indicated.

Classes of Atomic Plasma Emission Detectors

The Inductively Coupled Plasma (ICP). The ICP (*3*) is now the most widely used spectrochemical source in general analytical emission spectroscopy. The discharge results from the interaction of a radio-frequency field, at 27 or 41 MHz, with argon flowing through a quartz tube set within a copper induction coil. A spectrally intense stable plasma is produced with temperatures which can attain 9-10,000 K. Samples are introduced as gases, liquids or powdered solids. The usual arrangement uses a spray chamber - nebulizer combination similar to that used in flame spectroscopy. All compounds are considered to be completely atomised by this high temperature plasma and chemical and molecular interferences should be negligible. The ICP discharge is a well suited for liquid chromatographic detection since it is usually configured for a liquid inlet stream.

The Direct-Current Plasma Jet (DCP). DC exitation sources for atomic emission typically involve a low-voltage (10-50V), high-current (1-35 A) discharge. The DC plasma induced atomization and excitation produces a line-rich spectrum which makes it valuable for multi-element qualitative and quantitative analysis. Application of DC discharges for liquid samples is limited, but the direct-current plasma jet (DCP) discharge stabilized by flowing inert gas, usually argon, is better (4). A cathode jet is placed above two symetrically placed anode jets in an inverted 'Y' configuration; flowing argon causes vortexes around the anodes and a 'thermal pinch' gives an arc column of high current density and temperature. Solutions are introduced from nebulizer-spray chambers upwards into the junction of the two columns, where analyte spectral emission is observed.

The argon based ICP and DCP have found their major chromatographic interfacing applications for HPLC (including ICP-Mass Spectroscopy) and SFC for specific determination of the elements, primarily metals which are readily excited in argon plasmas.

The Microwave Induced Plasma (MIP). The plasma most used for GC detection is the microwave induced plasma (MIP). A helium or sometimes argon plasma is maintained within a 'cavity' which serves to focus power from a microwave source (usually operated at 2.45 GHz) into a discharge cell, (often a quartz capillary tube). Microwave plasmas may be operated at atmospheric or reduced pressures (5,6). Operational power levels of 50-150 watts are much lower than for DCP (ca. 500 watts) or ICP (1 - 5 kilowatts), making their operation less complex and inert gas consumption much lower. Power densities are similar, however, due to the small size of the microwave induced plasmas. Although plasma temperatures are lower in the helium MIP, the unique chemistry of helium discharges produces high spectral emission intensities for many elements including non-metals. Generally, non-metals respond poorly in the argon ICP or DCP. Low power MIP systems have been found less useful for liquid introduction since there is usually insufficient plasma enthalpy to desolvate and vaporize aerosols effectively; this problem leads to plasma instability and extinction.

The cavity which has been most used for GC detection has been the atmospheric pressure TM_{010} cylindrical resonance cavity developed by Beenakker (7) and its modifications such as the reentrant cavity depicted in Fig. 1 (8). A valuable feature is that emitted light is viewed axially, in contrast to 'transverse viewing' when observation must be made through cavity walls whose properties can change with time. The advantages inherent in atmospheric pressure operation greatly simplify GC detection.

The Surface Wave Plasma (SWP). An alternative to the electromagnetic resonant cavity for the transmission of microwave energy to the plasma is the surface wave induced plasma (SWP) using a 'surfatron' power launching device (9). Here the electric field applied to the plasma comes from a wave travelling along the plasma column and the device produces stable plasmas at different frequencies at reduced or atmospheric pressures.

Other Plasma Detectors. A 60Hz alternating-current helium plasma (ACP) has been used for chromatographic detection (10). It is a stable, self-seeding emission source, requiring no external initiation which does not extinguish under high solvent

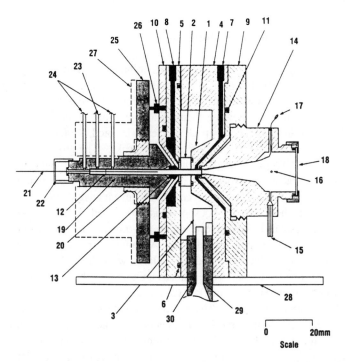

Figure 1. Reentrant Cavity: (1) pedestal, (2) quartz jacket, (3,4) main cavity body, (5) cavity cover plate, (6) gasket, (7,8) cooling water inlet and outlet, (9,10) water plates, (11) O-ring, (12) silica discharge tube, (13) polyimide ferrule, (14) exit chamber, (15,16) window purge inlet and outlet, (17) sparker wire, (18) window, (19) gas union, (20) threaded collar, (21) column, (22) capillary column fitting, (23) makeup and reagent gas inlet, (24) purge flow inlets, (25) stainless steel plate, (26) standoff, (27) heater block, (28) mounting flange, (29) brass center conductor, (30) PTFE coaxial insulator. (Reproduced from ref. 8. Copyright 1990 American Chemical Society.)

loads. Capacitively coupled atmospheric pressure plasmas (CCP) have proved effective for capillary GC without appreciable band broadening (*11,12*) and a development, the stabilized capactive plasma detector (SCP) has been introduced commercially (*13*). Radio frequency plasma detectors have been shown effective for both GC and SFC (*14*) High voltage electrode-less discharges in argon,nitrogen or helium, have been used for GC, utilizing emission from the atmospheric pressure afterglow. The metastable energy carriers in the helium system have the highest collisional energy transfer and the best ability to excite other elements to emission (*15*).

The Plasma - Chromatograph Interface.

Gas Chromatographs. Since eluent from GC columns is normally at atmospheric pressure, simpler interfacing configurations are possible for atmospheric pressure plasmas than for reduced pressure plasmas. Interfacing reduced pressure plasmas with GC involves evacuating an interface chamber to a pressure of about one torr. With packed columns, little degradation in peak efficiency is noted, but the interface volume leads to some inevitable broadening of capillary peaks. The atmospheric pressure plasmas such as the TM_{010} MIP are simple to interface with capillary GC columns since the latter can be terminated within a few millimeters of the plasma, giving only microliters of 'dead volume'. Heating is needed to prevent analyte condensation along the interface line and both cold and hot spots must be avoided. Helium make-up gas or other reactant gases can be introduced within the transfer line to optimize plasma performance and minimize peak broadening. Advantages in the performance of the GC-MIP are obtainable with a threaded tangential flow torch (TFT) (*16,17*), to give a self-centering plasma which can give enhanced emission and better stability, but high volumes (liters per minute) of helium flow gas are needed.

   Liquid Chromatographs. The HPLC interface to the ICP or the DCP is straightforward, since a heated transfer line is not usually needed, but it is important to reduce post-column peak broadening by minimizing the interface tube length and volume.   The major sensitivity limit of these HPLC interfaces is the relatively poor (1% or less) transfer efficiency of eluent into the plasma excitation region due to ineffective nebulization and desolvation. A better method for HPLC-plasma interfacing uses a 'direct injection nebulizer' (*18*), which can transfer mobile phase flows of up to 0.5 ml/min into the plasma, without appreciable peak broadening, and with efficiency approaching 100% . Thermospray sample introduction has also given substantially improved transfer efficiencies in HPLC-ICP for both aqueous and nonaqueous systems (*19*).

   HPLC-MIP interfacing has proven more problematic since the low powered MIP cannot be directly interfaced to conventional HPLC columns, as the discharge is quenched by mL/min liquid flow streams Some approaches to this problem have been explored. A mixed gas oxygen-argon MIP was applied successfully for HPLC of mercury compounds. Methanol/ water mixtures with up to 90% of the former solvent were tolerated; detection limits for organically-bound Hg were in the ng range, but response was found to be dependent upon molecular structure (*20*). Zhang et al.

devised a moving-wheel sample transport-desolvation system in which aqueous solvent is evaporated by a flow of hot nitrogen, leaving dry analyte to be transported into the plasma, where it is volatilized, atomized and excited. The plasma used was a small volume 100 watt helium MIP, and detection limits were in the range 0.4 -20 μg of halogen (*21*). The top and side views of the LC-MIP interface are shown in Fig 2. In general the interfacing problems for lower-powered MIPs parallel those in HPLC-MS and HPLC-FTIR. Capillary HPLC columns with mobile phase flow rates of a few mL/min provide a possibility for helium MIP interfacing, but sample capacity is very limited.

Supercritical Fluid Chromatographs. SFC detection may be either prior to decompression, or at atmospheric pressure after a decompression stage,.the former mode employing mainly LC-like and the latter GC-like detectors. As for HPLC, the need in SFC plasma detection is to minimize post-column band broadening, here during the decompression stage. Fortunately the nebulizer-spray chamber systems often needed in HPLC interfaces are not required since, the fluid, as it leaves the column and decompression restrictor, becomes a gas at atmospheric pressure and can in principle transport all of the analyte peak in a readily atomized form. Successful capillary SFC interfaces have been devised for ICP using direct injection (*22*), with a SWP (*23*) and with a Beenakker MIP using a toroidal plasma sustained in an ICP type torch with auxiliary argon flow gas.(*24*)

Detection Parameters.

Interelement Selectivity Detection of the desired element with minimum interference from other elements present in the plasma is essential. Selectivity depends on the emission spectrum of the element and of possible interfering elements, and on the properties of the spectroscopic measurement system. A standard measure of inter-element selectivity is the peak area response per mole of analyte element divided by the peak area response of the 'background' element per mole of that element, measured at the emission wavelength. Selectivity against carbon is usually most important, but other background matrixes have their own selectivity criteria. Selectivities vary among plasmas and with instrumental conditions, so calibration is essential.

Elemental Sensitivity and Limits of Detection. Elemental sensitivity depends on the emission intensity at the wavelength measured . Each element has a number of possible spectral lines or bands for determination and the best must be chosen, taking into account selectivity criteria as well. Different emission lines have different sensitivities in different plasmas. Sensitivity, defined by the slope of the response curve, is less often used in C-AES than 'detection limit', expressed as absolute values of element mass (in a resolved peak) or in mass flow rate units. Detection limits for different elements vary greatly and this factor affects inter-element selectivity if appreciable spectral overlap is present.

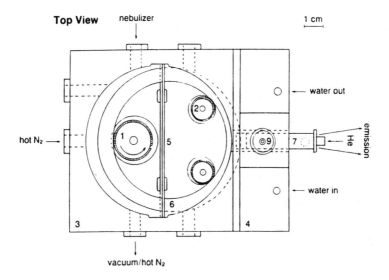

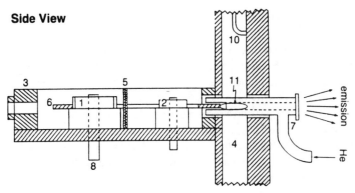

Figure 2. LC-MIP Interface : 1 is a friction wheel drive, 2 a guided wheel gearing, 3 the interface chamber housing, 4 the TM_{010} resonator, 5 the separation plate, 6 a stain-less steel wheel, 7 the fused silica plasma torch with face plate, 8 the driver wheel shaft, 9 an N connector, 10 the coupling loop, and 11 the plasma region. (Reproduced from ref. 21. Copyright 1989 American Chemical Society.)

Gas Chromatographic Applications.

GC-MIP Detection - Non-metals. The helium MIP has been the most used for non-metals detection, since for many such elements, argon plasmas do not provide efficient collision energy transfer for adequate excitation.

Reduced Pressure GC-MIP. McCormack et al. (*25*) and Bache and Lisk (*26*) reported reduced pressure argon GC-MIP for P, S, F, Cl Br, I, and C, with detection limits between 10^{-7} and 10^{-12} g/s, but selectivities against carbon were very poor. McLean et al. described a tunable multi-element detection system with detection limits between 0.03 and 0.09 ng/s for C, H, D, F, Cl, Br, I and S, and selectivities from 400 to 2300 (*6*). As and Sb were determined in environmental samples by derivatization with detection limits of 20 and 50pg (*27*). A commercial multi-channel instrument using a polychromator and Rowland Circle optics was evaluated by Brenner (*28*). For operation at 0.5 - 3 torr, with various scavenge gases, detection limits ranged from 0.02 ng/s for Br to 4 ng/s for O. Linearity was 3-4 decades and selectivities compared to carbon were 500-1000. Hagen et al. (*29*) used elemental derivatization taggants such as chlorofluoroacetic anhydride to introduce Cl and F for specific detection of acylated amines, while Zeng et al. (*30*) optimized oxygen specific detection. Sklarew et al.(*31*) noted that although the low pressure plasma gives optimal excitation energy for non-metals, its sensitivity is poorer than that of TM_{010} cavities because of its transverse viewing geometry. Olsen et al. (*32*) compared reduced pressure and atmospheric pressure MIPs for Hg, Se and As detection in shale oil matrices, and found the latter to give better detection limits and selectivities.

Atmospheric Pressure GC-MIP. Increased efficiency of transfer of microwave power to the discharge using cavity structures such as the Beenakker TM_{010}, allows plasmas to be sustained at atmospheric pressure. Another advantage is that light emitted from the plasma can be viewed axially, rather than transversely through the discharge tube wall,thus avoiding variable response upon extended tube use. The first applications of the cavity for packed column GC required solvent venting to avoid extinguishing the plasma (*33*). Flexible fused silica capillary GC columns can be terminated within a few mm of the plasma, and they are now universally used. Detection limits and selectivities from such systems are listed in Table 1 and widespread adoption of the technique for more routine use is now possible with the introduction of a commercial instrumental system which uses the cavity system depicted in Fig 1 (*8*). A rapid-scanning spectrometer has been applied for multi-element detection, elemental response per mole for C, Cl and Br being independent of molecular structure despite the low power (50-60 W) used (*34*). Near infra-red (NIR) atomic emission has been evaluated for GC-MIP, a cooled TM_{010} cavity and tangential flow torch being used at 370 watts with a Fourier transform NIR spectrometer (*35*).

Oxygen-selective detection has been reported by Bradley and Carnahan (*36*) with a TM_{010} cavity in a polychromator system. Background oxygen spectral emission from

**Table I.   Selected GC Detection Figures of Merit for Atmospheric
Pressure Helium MIP**

Element	λ (nm)	LOD pg/s (pg)	Selectivity (vs. C)	LDR
Carbon (a)	247.9	12.7 (12)	1	ca.1,000
Carbon (b)	193.1	2.6	1	21,000
Nitrogen (b)	174.2	7.0	6,000	43,000
Hydrogen (b)	486.1	2.2	variable	6,000
Sulfur (b)	180.7	1.7	150,000	20,000
Chlorine (b)	479.5	39	25,000	20,000
Oxygen (b)	777.2	75	25,000	4,000
Fluorine (b)	685.6	40	30,000	2,000
Silicon (b)	251.6	7.0	90,000	40,000
Germanium (a)	265.1	1.3 (3.9)	7,600	ca.1,000
Tin (a)	284.0	1.6 (6.1)	36,000	ca.1,000
Tin (b)	303.1	(0.5)	30,000	ca.1,000
Mercury (b)	253.7	0.1	3,000,000	ca.1,000
Lead (a)	283.3	0.17 (0.71)	25,000	ca.1,000
Iron (a)	259.9	0.3 (0.9)	280,000	ca.1,000
Vanadium (b)	268.8	10 (26)	57,000	ca.1,000
Nickel (b)	231.6	2.6 (5.9)	6,500	ca.1,000
Manganese (b)	257.6	1.6 (7.7)	110,000	ca.1,000

LOD- Limit of detection 3 times the signal to noise ratio
LDR- Linear dynamic range
(a) Conventional TM_{010} MIP - University of Massachusetts
(b) Cooled Torch, Hewlett Packard 5921A - from reference 8 or University of
Massachusetts

plasma gas impurities, leaks or back-diffusion into the plasma were minimized to give sensitivities between 2 and 500 ppm in different complex petroleum distillates. The selective detection of phenols in a light coal liquid distillate is seen in Fig 3, in which 2a shows carbon emission, 2b oxygen emission and 2c flame ionization detection for a phenolics concentrate of the same distillate.

TM_{010} cavities have been used for numerous capillary GC-MIP elemental investigations, including boron specific detection for diol boronate esters and other compounds (37). An current application of the reentrant cavity system for boron detection is shown in Fig. 4 which depicts boron detected as 1,2-hexadecyldiol phenylboronate in motor oil (38). The surfatron-MIP has proven to be of value in determinations of P, S, Cl and Br in pesticides. Detection limits ranged from 3 to 60 pg/s (39)

GC-MIP Detection - Metals. Many volatile organometallic and metal chelate compounds can be successfully gas chromatographed (40), and GC-AES detection is valuable in confirming elution of metal containing species, and acquiring sensitive analytical data. As shown in Table 1, GC-MIP data has been obtained for many metals, and some such as lead and mercury have been studied quite extensively; these elements are detectable with TM_{010} cavities to sub-pg/second limits. In their comparison of reduced and atmospheric pressure MIP systems, Olsen et al. found for the latter system a one pg detection limit for mercury, with selectivity over carbon of 10,000 (32). An early GC-MIP study of volatile elemental hydrides of germanium, selenium and tin gave sub-ng detection (41) and there is considerable potential for the determination of these elements in environmental matrixes using hydride forming derivatization techniques.

The GC of metal chelates of sufficient volatility and thermal stability for quantitative elution has widely studied, most applications being for complexing ligands of 2,4-pentanedione (acetylacetone) and its analogs (40). Among examples have been analyzed using GC-MIP is chromium as its trifluoracetylacetonate in blood plasma. Excellent quantitation and precision was obtained(42). Trace determinations of beryllium, copper and aluminum as volatile chelates have been reported and ligand redistribution and reaction kinetics of gallium, indium and aluminum chelates have been followed (43). Figure 5 illustrates the element specific detection, using the reentrant cavity system, of copper, nickel, and platinum chelates of N,N'-ethylene-bis(5,5-dimethyl-4-oxohexane-2-imine) (H_2enAPM_2) which forms stable complexes with divalent transition metals. The close proximity of emission lines for nickel and palladium is responsible for the small negative platinum signal seen for the nickel complex; refinement of the measurement 'recipe' may be able to eliminate this effect.

GC-DCP Detection. GC-DCP was used for the determination of methylcyclopentadienylmanganese tricarbonyl in gasoline. Direct injection of 5 μL of gasoline allowied determination in the low ppm range with a relative standard deviation of 0.8-3.4% (44). Redistribution reactions of silicon, germanium, tin and lead compounds were examined. Detection limits (pg/s) and carbon selectivities using a three-electrode plasma jet for these elements were: Cr (4, 4×10^8), Sn (60, 2.5×10^6),

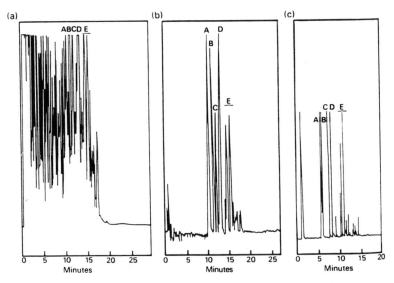

Figure 3.   Gas chromatograms of light coal distillates: a) Carbon emission, b) Oxygen emission, c) FID trace of phenolics concentrate. Peak A is phenol, peak B, o-chlorophenol, peak C, o-chlorocresol, peak D, m- and p-cresols and peak E, C2-phenols. (reference 36).

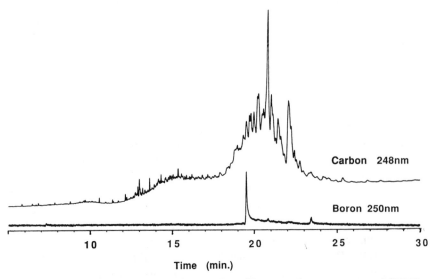

Figure 4. Simultaneously acquired B and C capillary gas chromatograms of 5W-30 motor oil spiked with 0.02 wt. % boron as 1,2hexadecyldiol phenylboronate. Column 25 m × 0.32 mm i.d. HP-1. 1ml injection split 80:1, column programmed from 60–270°C. (Reproduced with permission from ref. 38. Copyright 1991 Royal Society of Chemistry.)

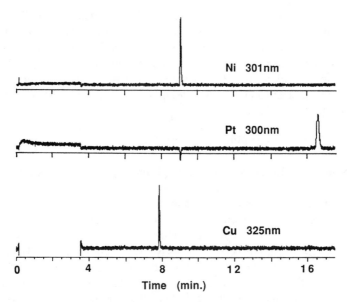

Figure 5.   Simultaneously acquired Cu, Ni and Pt specific capillary gas chromatograms of chelates of N,N'-ethylenebis(5,5-dimethyl-4-oxohexane 2-imine. Column 30 m x 0.25 mm i.d. DB-1. 250°C.

Pb (100, 5x10^5), and B (3, 3x10^5)10 (45). The versatility of GC-DCP has been exploited in a simple system dedicated to the specific determination of methylmercury compounds in fish (46)..

GC-SWP Detection. Surface wave plasmas typically are stable, reproducible and versatile in operation.(9). GC applications have been reported by a number of authors and an important example of their application is in multiple wavelength detection utilizing Fourier transform spectrometry.(47). Figure 6 shows an example of 'spectrochromatograms' of chlorofluoroethers using the spectral range 680 - 1080 nm to allow simultaneous detection of F, O, Cl and C.

GC-CCP Detection. Capacitively coupled plasmas have proved capable of reproducible and sensitive atomic emission detection. In contrast to the MIP, CCP systems utilize radiofrequency energy. A system designated the SCP (stabilized capacitive plasma) uses a radiofrequency at 27.12 MHz and up to 200 watts is capacitively coupled to the plasma. Although helium has been most used as the plasma gas because of its high excitation energy for non-metals such as halogens, in principle every gas would work for the SCP even at atmospheric pressure. Figure 7 depicts a multi-element chromatogram for pesticides measuring simultaneously C, S, Cl and Br in the 650 - 1000 nm region. (48).

Liquid Chromatographic Applications.

Development in HPLC-AES detection has been mainly with the ICP and to some extent with the DCP and the MIP. Metal specific detection predominates and will probably remain so until interface systems to remove HPLC mobile phases, while transfering eluate peaks to a plasma optimized for non-metals, can be devised. Such an interface can involve a moving band eluate transport device, or be based on thermospray (19) or particle beam technology. The major problem in HPLC-plasma interfacing is poor plasma compatibility with required mobile phase flow rates. Traditional specific element atomic spectroscopic detectors use on-line nebulization and excitation of small volumes (5 - 200 μL) of liquid which are converted into an aerosol and introduced into the atomization-excitation cell. The major reason for poor detection limits is the relatively ineffective conversion of effluent flow into aerosol and its transport to the plasma. Typically only 1 - 5% of the sample reaches the plasma. Also, there is poor tolerance of the plasma to common solvents used in HPLC. This is true particularly for reverse phase ion pairing and size exclusion chromatography. These problems need to be addressed by more quantitative nebulization, atomization and excitation of HPLC samples as well as better transport systems.

HPLC-ICP Detection. Many applications of HPLC-ICP have appeared since 1979 (2), but detection limits at levels of environmental significance have only been moderate. Detection limits for aqueous mobile phases, whose characteristics are familiar from standard sampling procedures, are usually two or more orders of magnitude worse than in continuous flow ICP-AES. Normal phase HPLC, using organic solvents such as hexane or methyl isobutyl ketone, presents further problems, since plasma behavior is less well defined and spectral background interference more

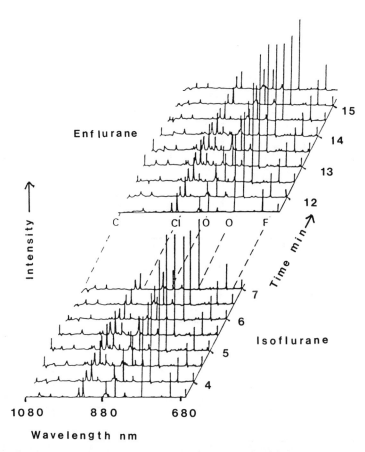

Figure 6. Spectro-chromatograms of isoflurane ($CF_3CHClOCHF_2$) and enflurane (CHF_2OCF_2CHFCl). Spectral resolution 16 cm^{-1}, elapsed time between spectra, 0.58 s. (Reproduced with permission from ref. 47. Copyright 1988 Royal Society of Chemistry.)

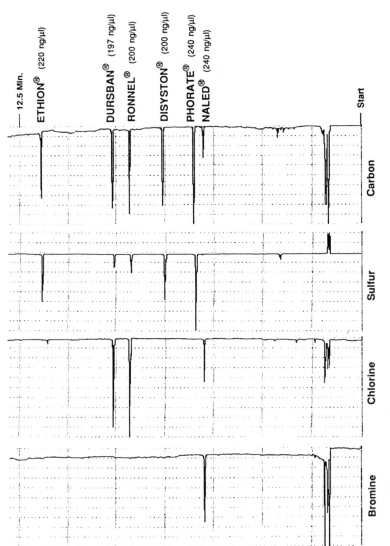

Figure 7. Comparison of stabilized capacative plasma detection at four elemental channels (Cl, Br, S and C). Mixture of 6 pesticides. Temperature program 1 min at 130°C, ramp 10°C/min to 260°C. (Reproduced with permission from ref. 48. Copyright 1990 Gordon and Breach.)

pronounced. One approach has been to use microbore HPLC to utilize lower mobile phase flow rates. For samples of copper and zinc diketonates and dithiocarbamates, peak broadening was minimised by optimal design of the interface, connecting tubing, nebulizer, spray chamber and plasma torch to give virtually identical peak width ratios for ICP and UV detection (*49*). At flow rates above 15 µL/min, sensitivity was independent of flow rate and the ICP operated as a mass flow sensitive detector; below that flow rate, however, it behaved as a concentration sensitive detector. HPLC-ICP has been effective for metalloid detection; a 130 ng/mL detection limit for arsenic in organo-arsenic acids being obtained for 100 µL samples (*50*). The ICP spectrometer utilized 48 channels operating at 1.2 kW, allowing sampled chromatograms for As, Se and P to be displayed on-line. Arsenic and cadmium compounds gave detection limits for 50µL injections of 3.1 ng/mL for As as arsenite and 0.12 ng/mL for Cd as the nitrilo-triacetate (*51*). Size exclusion (SEC) - ICP has given molecular size distribution of sulfur, vanadium and nickel petroporphyrin compounds in petroleum crudes and residuals (*52*.) Ferritin, an iron-containing protein existing in a number of forms, has been analyzed by aqueous SEC with good repeatability at ng levels (*53*).

Current developments suggest that notable enhancement in working sensitivities is feasible. To overcome transfer problems of HPLC eluate to the ICP, a useful device is the Direct injection nebulizer (DIN) (*18*), a total injection microconcentric nebulizer which can achieve almost 100% nebulization and transport efficiency. Examples of detection limits were 164 ng/mL for sulfur and 4 ng/mL for zinc. Thermospray sample introduction has given enhancements in detectability of ca. 30 over pneumatic nebulization. Figure 8 shows separation of Cr(III) and Cr(VI) by ion-pair chromatograhy using pentanesulfonic acid (*19*). Developments in HPLC-ICP-MS are considered later.

HPLC-DCP Detection. The good tolerance of the DCP to a wide range of solvents has shown its value as an HPLC detector. Standard nebulization was used for reverse phase chromatography, but an impact device proved better for normal phase hydrocarbon and halocarbon eluents. Metal chelates showed mass flow detection limits of 0.3 ng/s for copper and 1.25 ng/s for chromium (*54*). The determination of Cr(III) and Cr(VI) (as chromate) by reverse phase ion pairing gave detection in the 5-15 ppb range (*55*). Applications included biological samples from ocean floor drillings, chemical dump sites, surface well water and waste water samples.

HPLC-MIP Detection. A direct introduction system into an MIP involved flowing the liquid effluent over a heated wire and vaporizing it by a cross-stream of helium into the discharge (*56*). The moving-wheel sample transport-desolvation system described earlier shows promise with acceptable detection limits, but the device is mechanically complex. The direct injection nebulizer, thermospray or particle beam approaches may also prove useful for HPLC-MIP Another possibility is cryo-focusing as also used in HPLC-FTIR. Investigations are also underway on new plasma cavities which may be able to sustain the helium MIP under conventional HPLC flow conditions (*57*).

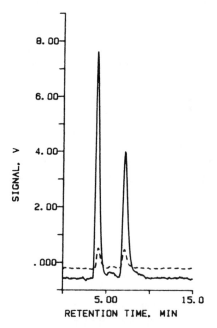

Figure 8. Separation of Cr(III) and Cr(VI) by ion pair chromatography with 50 μm thermospray (—) and pneumatic (----) sample introduction. (Reproduced from ref. 19. Copyright 1990 American Chemical Society.)

Supercritical Fluid Chromatographic Applications.

Analytical SFC has become accessible through the availability of high resolution packed and capillary SFC columns and instrumentation. High resolution SFC along with supercritical extraction, may allow separations in areas where neither GC or HPLC are possible. A SWP sustained in helium was used for SFC, giving sulfur-specific detection at 921.3 nm with a 25 pg/s limit for thiophene (23).An initial report described an ICP interface with close to 100% atomization efficiency (58), and a practical interface has been devised for SFC-ICP and applied to organosilicon compound separations (22). A radio frequency plasma (RFP) gave good performance for S and Cl detection at 921.3 and 837.6 nm respectively, and minimal spectral interferences were seen for carbon dioxide and nitrous oxide doped plasmas. Detection limits ranged from 50-300 pg/s. Figure 9 shows comparative SFC chromatograms for FID and RFP (Cl) detection for DDT in a milk extract (14). Modification of plasma excitation by SFC solvents appears to be less difficult than for typical organic HPLC solvents, and it seems likely that as SFC becomes more widely adopted, element specific detection by atomic plasma emission will become a useful option.

Chromatographic Applications of Plasma - Mass Spectrometry.

Plasma source mass spectrometry, in which the plasma generates ions for mass spectral identification and quantitation, has become established as a major analytical technique particularly in the case of ICP-Mass Spectrometry (59) . Other plasmas are also applicable as mass spectral ion sources.

HPLC - Plasma Mass Spectrometry. Among the most sophisticated developments in plasma detection for chromatography has been the development of 'double interfaced' systems such as HPLC-ICP-Mass Spectrometry (60) for which detection limits of 500 pg/peak or below have been found for many elements. Ion exchange and ion pair chromatography were used for triorganotin species (61), and arsenic speciation has been the subject of a number of studies also using ion exchange (62,63) .This technique shows much potential in biomedical and clinical studies where analyte levels are usually below the capabilities of ICP emission detection. Gercken and Barnes (64) employed aqueous size exclusion chromatography (SEC) ICPMS for on line element and isotope ratio detection of lead and copper in protein fractions of serum and red blood cell hemolysate from >600 to 11 kilodaltons. Figure 10 shows distributions of lead, copper and zinc in rat blood serum. Heitkemper and co-workers have coupled HPLC to MIP-MS for the determination of halogenated organic compounds (65). Bromine and iodine were determined at low pg levels but chlorine determination was hindered by large background signals.

GC - Plasma Mass Spectrometry. Efforts to develop interfaced GC methods have been relatively limited.since GC-ICP itself has attracted little attention. However, coupling packed column GC with ICP-MS has been reported to give detection limits frm 0.001 to 400 ng/s for a range of elements including halogens, silicon, phosphorus and sulfur (66). The most promising approach is that of GC-MIP-MS, as yet

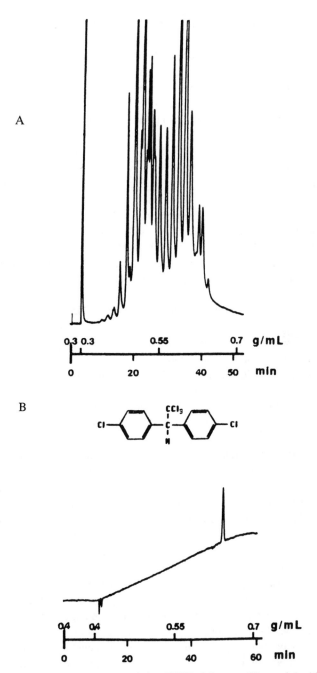

Figure 9. SFC of a milk extract containing DDT. 2.5 m × 50 μm i.d. biphenyl capillary column; CO_2 at 100°C. (A) FID; (B) Cl detection at 837.6 nm. (Reproduced from ref. 14. Copyright 1989 American Chemical Society.)

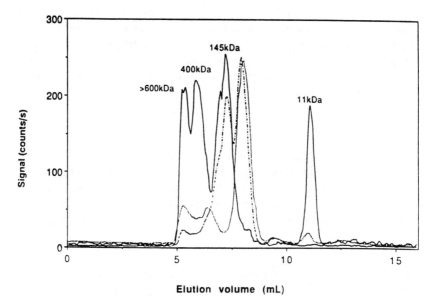

Figure 10: SEC-ICP-MS chromatogram of ^{208}Pb (solid line), ^{63}Cu (–--–), and ^{64}Zn (dotted line) distribution in rat blood serum. Evaluated molecular weights of lead fractions are labeled. (Reproduced from ref. 64. Copyright 1991 American Chemical Society.)

commercially undeveloped, but capable of extending the positive features of GC-MIP to better detection limits (67).

Conclusions.

The instrumental and experimental requirements for interfaced or tandem analytical techniques are indeed challenging when the goals are to accomodate a wide range of separation methods, to achieve a high level of confidence in elemental speciation, and to attain the greatest possible sensitivity and versatility. The contributions of many research workers cannot be adequately acknowledged. Separation scientists have posed the problems and analytical spectroscopists have answered them in large measure. The field of endeavor has developed very strongly over the past two decades and it is the goal of this symposium volume to present a state of the art coverage of much of the broad area of combination methods incorporating analytical separations and atomic spectroscopy. There are still many goals to be met but the potential for continued success is very strong.

Acknowledgements.

I thank the many colleagues and students who have made this area of research so worthwhile and productive over the past 15 years or more. Much remains to be done, relying on their continued enthusiasm. In particular I salute my colleague Ramon Barnes for continually stressing spectroscopic rigor to temper chromatographic enthusiasm, and Bruce Quimby for carrying doctoral research to commercial fruition.

Literature Cited

(1)   Ebdon, L.; Hill, S.; Ward, R. W., *Analyst,* **1986,** *111,* 1113.
(2)   Ebdon, L.; Hill S.; Ward, R. W., *Analyst,* **1987,** *112,* 1.
(3)   Barnes, R. M., *Crit.Rev.in Anal.Chem.,* **1979,** *7,* 203.
(4)   Decker, R. J., *Spectrochim.Acta,* **1980,** *35B,* 19.
(5)   Bache C. A.; Lisk, D. J, *Anal.Chem.,* **1977,** *39,* 786.
(6)   McLean, W. R.; Stanton, D. L.; Penketh, G. E., *Analyst,* **1973,**98, 432.
(7)   Beenakker, C. I. M. *Spectrochim. Acta* **1977,** *32B,* 173.
(8)   Quimby, B. D.; Sullivan, J. J. *Anal. Chem.* **1990,** *62,* 1027.
(9)   Hubert, J.; Moisan, M.; Ricard, A., *Spectrochim. Acta.,* **1979,** *34B,* 1.
(10)   Costanzo, R. B.; Barry, E. F., *Anal. Chem.,* **1988,** *60,* 826.
(11)   Patel, B. M.; Heithmar, E.; Winefordner, J. D., *Anal. Chem.,* **1987,** *59,* 2374
(12)   Liang, D. C.; Blades, M. W., *Anal. Chem.,* **1988,** *60,* 27.
(13)   Anton Paar Company, Graz, Austria.
(14)   Skelton, R. J. Jr.; Farnsworth, P. B.; Markides, K. E.; Lee, M. L., *Anal. Chem.,* **1989,** *61,* 1815.
(15)   Rice, G. W.; D'Silva, A. P.; Fassel, V. A., *Spectrochim. Acta.,* **1985,** *40B,* 1573.
(16)   Bollo-Karmara; Codding, E. G., *Spectrochim. Acta* **1981,** *36B,* (10), 973.
(17)   Goode, S. R.; Chambers, B.; Buddin, N. P. *Spectrochim. Acta* **1985,** *40B(1/2),* 329.

(18)   LeFreniere, K. E.; Fassel, V. A.; Eckel, D. E., *Anal. Chem.*, **1986**, *59*, 879.
(19)   Roychowdhury, S. B.; Koropchak, J. A., *Anal. Chem.*, **1990**, *62*, 484.
(20)   Kollotzek, D.; Oechsle, D.; Kaiser, G.; Tschopel P.; Tolg, G.,*Fresenius Z.Anal.Chem.*, **1984**, *318*, 485.
(21)   Zhang, L.; Carnahan, J. W.; Winans, R. E.; Neill, P. H., *Anal. Chem.*, **1989**, *61*, 895.
(22)   Forbes, K. A.; Vecchiarelli, J. F.; Uden, P.C.,Barnes, R. M., *Anal. Chem.* **1990**,*.62*, 2033
(23)   Lufffer, D. R.; Galante, L. J.; David, P. A.; Novotny, M.; Hieftje, G. M., *Anal. Chem.*, **1988**, *60*, 1365.
(24)   Motley, C. B.; Ashraf-Khorassani, M.; Long, G. L., *Applied Spectrosc.* **1989**, *43*, 737.
(25)   McCormack, A. J.; Tong S. C.; Cooke, W. D., *Anal.Chem*, **1965**, *37*, 1470.
(26)   Bache C. A.; Lisk, D. J., *Anal.Chem.,.*1965, *37*, 1477.
(27)   Talmi, Y.; Norvall, V. E., *Anal. Chem.*, **1975**, *47*, 1510.
(28)   Brenner, K. S., *J. Chromatogr.*, **1978**, *167*, 365.
(29)   Hagen, D. F.; Marhevka, J. S.; Haddad, L. C., *Spectrochim Acta*, **1985**, *40B*, 335.
(30)   Zeng, K.; Gu, Q.; Wang G.; Yu, W.; *Spectrochim Acta,.*1985, *40B*, 349.
(31)   Sklarew, D. S.; Olsen, K. B.; Evans, J. C., *Chromatographia*, **1989**, *27*, 44.
(32)   Olsen, K. B.; Sklarew, D. S.; Evans, J. C.,*Spectrochim Acta*, **1985**, *40B*, 357.
(33)   Quimby, B. D.; Uden, P. C.; Barnes, R. M. *Anal.Chem,.*1978, *50*, 2112.
(34)   Zerezghi, M.; Mulligan, K. J.; Caruso, J. A., *J. Chromatog. Sci.*, **1984**, *22*, 348.
(35)   Pivonka, D. E.; Fateley, W.G.; Fry, R. C., *Appl. Spectrosc.*, **1986**, *40*, 291.
(36)   Bradley, C.; Carnahan, J. W., *Anal. Chem.*, **1988**, *60*, 858.
(37)   Sarto, L. G. Jr.; Estes, S. A.; Uden, P. C.; Siggia, S.; Barnes, R. M. *Anal. Letters* **1981**, *14*, 205.
(38)   Seeley, J. A.; Uden, P. C., *Analyst*, **1991**, to be published.
(39)   Riviere, B.; Mermet, J-M.; Deruaz, D., *J. Anal. Atomic. Spect.*, **1987**, *2*, 705.
(40)   Uden, P. C., *J. Chromatogr.*, **1984**, *313*, 3.
(41)   Robbins, R. B.; Caruso, J. A., *J. Chromatogr. Sci.*, **1979**, *17*, 360.
(42)   Black, M. S.; Sievers, R. E., *Anal. Chem.*, **1976**, *48*, 1872.
(43)   Uden, P. C.; Wang. T., *J. Anal. Atomic. Spect.*, **1988**, *48*, 1872.
(44)   Uden, P. C.; Barnes, R. M.; DiSanzo, F. P., *Anal.Chem.*, **1978**,*50*, 852.
(45)   Estes, S. A.; Poirier, C. A.; Uden, P. C.; Barnes, R. M., *J.Chrom.*, **1980**, *196*, 265.
(46)   Panaro, K. W.; Erikson, D.;Krull, I. S., *Analyst*, **1987**, *112*, 1097.
(47)   Lauzon, C.; Tran, K. C.; Hubert, J., *J. Anal. Atomic. Spect.*, **1988**, *3*, 901.
(48)   Knapp, G.; Leitner, E.; Michaelis, M.; Platzer, B.; Schalk, A., *Intern. J. Environ. Anal. Chem.*, **1990**, *38*, 369.
(49)   Jinno, K.; Tsuchida, H.; Nakanishi, S.; Hirata, Y.; Fujimoto, C., *Appl.Spectrosc.*, **1983**, *37*, 258.
(50)   Irgolic, K. J.; RStockton, R. A.; Chakraborti, D.; Beyer, W., *Spectrochim.Acta*, **1983**, *38B*, 437.
(51)   Nisamaneepong, W.; Ibrahim, M.; Gilbert , T. W.; Caruso, J. A., *J.Chromatog.Sci.*, **1984**, *22*, 473.

(52)   Hausler, D. W., *Spectrochim. Acta.*, **1985**, *40B*, 389..

(53)   La Torre, F.;Violante, N.;Senofonte, O.; D'Arpino C.; Caroli, S., *Spectroscopy*, **1989**, *4(1)*, 48.

(54)   Uden, P. C.; Quimby, B. D.; Barnes, R. M.; Elliott,.W. G., *Anal Chim. Acta*, **1978**, *101*, 99.

(55)   Krull, I. S.; Panaro, K. W.; Gershman, L. L., *J.Chromatog.Sci.*, **1983**, *21*, 460.

(56)   Billiet, H. A. H.; J van Dalen, J. P.; Schoemakers, P. J.; deGalen, L., *Anal.Chem.*, **1983**, *55*, 847.

(57)   Forbes, K. A.; Reszke, E. E.; Uden, P. C.; Barnes, R. M., *J. Anal. Atomic.Spectr.*, **1991**, *6*, 57

(58)   Olesik, J. W.; Olesik, S. V., *Anal. Chem.*, **1987**, *59*, 796.

(59)   Houk, R. S., *Anal. Chem.*, **1986**, *58*, 97A.

(60)   Thompson, J. J.; Houk, R. S., *Anal.Chem.*, **1986**, *58*, 2541.

(61)   Suyani, H.; Creed, J.; Davidson, T.; Caruso, J. A., *J.Chromatog.Sci.*, **1989**, *27*, 139.

(62)   Beauchemin, D.; Bednas, M. E.; Berman, S. S.; McLaren, J. W.; Siu K. W. M.; Sturgeon, R. E.; *Anal.Chem.*, **1988**, *60*, 2209.

(63)   Heitkemper, D.; Creed, J.; Caruso,J. A.; Fricke, F.*J.Anal.Atomic.Spect.* **1989**, *4*, 279.

(64)   Gercken, B.; Barnes, R. M. *Anal. Chem.*, **1991**, *63*, 283

(65)   Heitkemper, D.; Creed, J.; Caruso, J. A. *J. Chromatog. Sci.*, **1990**, *28*, 175

(66)   Chong, N. S; Houk, R. S.*Appl. Spectrosc.***1987**, *41*, 66.

(67)   Satzger, R. D; Fricke, F. L; Brown, P. G.; Caruso. J. A. *Spectrochim Acta*, **1987**, *42B*, 705.

RECEIVED September 5, 1991

Chapter 2

Atomic Emission Spectrometry with Helium Plasmas for Liquid and Supercritical-Fluid Chromatography

Gregory K. Webster and Jon W. Carnahan

Department of Chemistry, Northern Illinois University, DeKalb, IL 60115

Nonmetal excitation capabilities have provided the impetus for numerous studies of the analytical behavior of helium plasmas. Techniques involving gas chromatography and helium plasmas have matured; initial results coupling helium plasmas with supercritical fluid and liquid chromatography are promising. An overview of the mating of the nonmetal detection capabilities of helium plasmas with liquid and supercritical fluid chromatography is presented. Important considerations discussed include solvent-plasma interactions, interface design, and spectral behavior.

With the evolution of laboratory technology and increasing demands upon the analyst, both the potential and the desire to analyze increasingly complex organic samples have increased. Generally, nonvolatile components of these samples must be separated by chromatographic techniques to ensure unambiguous detection and determination. Currently, the most popular of these techniques is liquid chromatography (LC). A second technique undergoing rapid development for the separation of high molecular weight compounds is supercritical fluid chromatography (SFC). No single detection system stands out for either separation technique which is easy to implement, possesses high sensitivity, exhibits response of a universal nature, has a high information content, and produces easily interpretable data. However, the aforementioned behaviors have been exhibited with helium plasmas when used as gas chromatography (GC) detectors (1-9). Coupling LC and SFC to helium plasmas is a logical progression from the successes of these GC-helium plasma systems. Initial successes have been demonstrated using helium plasmas as detectors for LC and SFC (10-17). In this paper, fundamental considerations regarding the potential, the current status and future directions of LC and SFC-helium plasma systems are discussed and evaluated.

0097–6156/92/0479–0025$06.00/0

Selective/Universal Detectors

One of the major drawbacks of LC is the limited choice of detectors. Although a few "selective" detectors exist for LC, the selectivity is not predictable, a priori. Conversely, "universal" detectors tend to be of relatively low sensitivity and response cannot rigorously be predicted without unique molecular information such as molar absorptivity, etc. The most commonly utilized LC detectors are based upon molecular absorbance, molecular fluorescence, or refractive index (RI); these work well for many analyses, but none possess both universal and "tunable" detection characteristics of high selectivity.

A much better situation exists for SFC, since many of the characteristics of the mobile phase are compatible with the broad array of GC detectors. In fact, GC type detector designs for SFC often incorporate mobile phase expansion from the supercritical fluid state to the gaseous state as an aid to analyte transport. The stalwart SFC detector has been the flame ionization detector (FID) which is universal but offers essentially no selectivity. Other than mass spectrometry, no SFC detector has emerged with "tunable" selectivity.

Very few detectors are simultaneously sensitive, selective and universal. When used for GC detection, atomic emission spectrometry (AES) with helium plasmas offers low detection limits and element selective detection for potentially every element of the periodic table, except helium. Metal detection limits as low as 0.28 pg/s have been reported for iron (6), and nonmetal detection limits as low as 0.6 pg/s for hydrogen (18). Estes, Uden and Barnes (6) obtained detection limits for nonmetals such as Br, C, Cl, I, and P in the 2.7 to 43 pg/s range. The capability of spectrally monitoring elements such as C and H with compound independent response, while monitoring hetero-atoms such as Cl and S, offers both "universal" and "selective" modes of detection with helium plasmas. Recently, multielement detection with helium plasmas has become commercially available for GC (9).

Liquid Chromatography with Helium Plasma Detection Systems

The use of helium plasma systems as GC detectors is straight forward, for helium is used as both the GC mobile phase and the plasma support gas. Significant effects on spectral and excitation behavior of these plasmas have not arisen from the mixing of these two helium streams. The first problem that presents itself when extending AES detection to other chromatographic techniques is the use of mobile phases which alter the characteristics of the helium plasma. Most studies, to date, dealing with LC-helium plasma detectors have a major focus on the interface, with mobile phase considerations being critical. In reports on LC-helium plasma systems for nonmetal determinations, analyte has been introduced into the plasma in three ways: direct routing of the LC effluent into the plasma region (10,11), nebulization of the effluent directly to the plasma (12,13) and methods which involve separation of the mobile phase from the analyte before introduction into the plasma (14).

Huf and Jansen (10,11) routed LC mobile phase directly to an atmospheric pressure helium plasma maintained with coaxial power coupling (Figure 1). This system was used to sustain a low power plasma at 75W for the LC application, an advantage being that a plasma containment tube was not necessary to confine the plasma. Thus, problems of carbon deposition, memory effects, or background atomic emission signals from torch degradation were not evident. This system worked well for GC eluates and analytes vaporized from thin layer chromatography plates.

Spectra obtained when methanol and chloroform were directed into the plasma showed chlorine atomic emission lines for the latter solvent. Unfortunately, this system was not analytically useful for LC detection due to a lack of sensitivity with direct effluent introduction. The authors suggested a thermal evaporation and atomization interface would be necessary. The problems encountered with this plasma exemplify the major obstacle encountered when directly coupling low power helium plasmas with LC. The low power plasma is not capable of desolvation, atomization and excitation of the analyte in the presence of the entire flow of the mobile phase (ca. 10 mg/s). Even trace amounts of foreign gases can cause significant signal reduction in the ability of low power plasma systems to excite nonmetals *(19)*. In contrast, a maximum of only a few micrograms of "foreign" material reaches the plasma at one time in gas chromatography. The low power plasma displays no problems with atomization and excitation under these conditions.

Billiet and coworkers *(12)* reported on deuterium selective detection of D_2O at 656.10 nm for fundamental studies of reversed-phase liquid chromatography behavior. Figure 2 is a schematic of the microwave-induced plasma (MIP) interface. Operationally, LC effluent was routed onto a wire and analyte with sufficient vapor pressure was carried into the plasma by flowing helium. In this manner, only a fraction of the volatile LC effluent was introduced into the plasma. This LC-MIP system and a RI detector were examined and compared for binary and ternary mobile phases. In general, the signal from the RI detector was more complex and difficult to interpret than the MIP response. The authors did report a loss in sensitivity in D_2O detection when tetrahydrofuran was included in the solvent mixture, attributable to increased carbon deposition in the quartz tube. This problem was reduced by cooling the plasma tube, frequently changing the plasma tube, and adding oxygen to the helium support gas.

The 500W MIP can tolerate increased sample loadings compared to the low power systems mentioned thus far. Using ultrasonic nebulization (USN), direct determination of nonmetals with ppm and sub-ppm detection limits has been reported for aqueous samples *(20-23)*. System figures of merit with LC effluent nebulization and helium plasma detection were reported by Michlewicz and Carnahan *(13)*. With this system, ion exchange chromatography eluates were ultrasonically nebulized into a moderate power (500W) MIP. This scheme is similar to those utilized with direct current and inductively coupled plasmas as LC detectors for metals *(24-34)*. It is important to note that Allen and Koropchak *(35)* have shown that the use of a nebulizer with LC effluent does not significantly contribute to band spreading.

Figure 3 illustrates the selective detector response at wavelengths specific for chlorine, bromine and iodine emission using USN-LC-MIP. The retention time difference for ClO^-, BrO_2^- and IO_3^- is only 0.4 minute, and with a nonselective detector a mixture of these species could not be determined. Calibration plots for this analysis were generally linear in a range from 10 to 100 μg for each nonmetal. Detection limits were of the order of 2-6 μg of

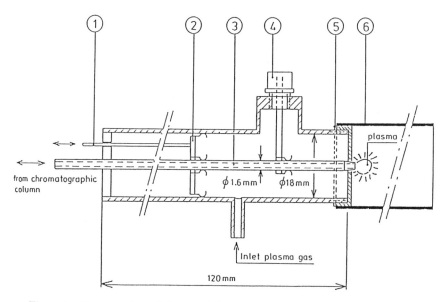

Figure 1. Cross-section of the coaxial plasma source used by Huf and Jansen. (1) Positioning rod of the piston with precision springs; (2) piston with precision springs; (3) capillary stainless-steel tube (1.6mm o.d., 0.5mm i.d.); (4) N-type connector; (5) exchangeable nozzle; (6) glass cylinder. (Reproduced with permission from ref. 11. Copyright 1985 Pergamon Press Ltd.)

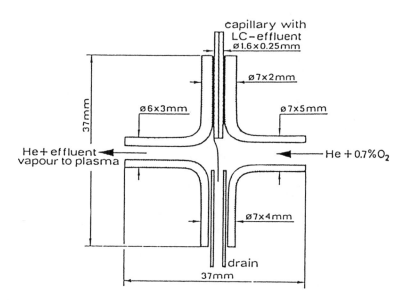

Figure 2. The interface between the LC apparatus and the MIP detector for Billiet et al. (Reproduced with permission from ref. 12. Copyright 1983 American Chemical Society)

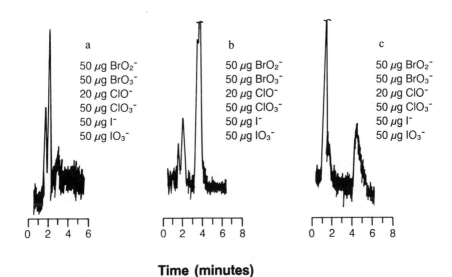

Time (minutes)

Figure 3. Separation of a six component mixture and response to non-analyte containing species with the USN-LC-MIP system. Monitored wavelength: a, 478.55nm Br(II); b, 479.45 nm Cl(II); c, 206.24 nm I(I). (Adapted with permission from ref. 13. Copyright 1987 Marcel Dekker, Inc.)

halogen (60-120 ng/s). Consistent with GC-MIP, response was proportional to the halogen mass and independent of the chemical form. Thus, changes in detector response due to acid-base character and conductivity of the analyte species were not of concern for MIP detection.

These USN-LC-MIP detection limits were governed by the intense spectral background produced upon the introduction of organic mobile phases, as seen in Figure 4. The 2.5% methanol mobile phase produced intense molecular band emission and spectral overlap problems not seen with aqueous samples *(36)*. This band emission can increase the noise at the wavelength of the analyte emission line and, in turn, degrade the detection limit.

The difficulties produced by allowing the LC mobile phase to reach the plasma were minimized with a moving wheel interface *(14)*, shown in Figure 5. The effluent, in the form of a mist, was directed from an anion exchange column to a rotating stainless steel wheel, from which the solvent was evaporated by a flow of hot nitrogen. Analyte residue on the wheel was transported to and directly contacted a low power (100W) helium-MIP for volatilization, decomposition and AES detection.

The rotating wheel interface had three advantages compared to the USN-LC-MIP system: (1) lower power requirement, (2) lower helium flow and (3) no spectral/excitation interference from the LC mobile phase. Analytes examined included oxo-halogen acids, which gave chromatograms very similar to those of Figure 3. Calibration plots were again linear from 10 to 100 μg and detection limits were in the 10 to 300 ng/s range. These detection limits were of the same magnitude as those obtained with USN-LC-MIP. Importantly, the limits of detection with the moving wheel interface were not spectral interference limited, but were governed by flicker noise created by interactions of the plasma with the wheel. With mechanical modifications, it may be possible to improve these detection limits, but because of the difficulty associated with controlling the temperature of the wheel, it is unlikely that this system will be widely useful for organic analysis. However, for analytes with boiling points much greater than the solvent, the moving wheel interface shows promise.

Concerning the current status of LC with helium plasma atomic emission detection, a few general conclusions may be made:

1. The presence of common LC organic mobile phases causes spectral and excitation interferences in the helium plasma.
2. Lower power plasmas, while capable of nonmetal excitation, are especially susceptible to extinguishment and reduction of nonmetal emission intensity by the introduction of LC solvents. To make these systems operable as LC detectors, removal of solvent prior to analyte introduction is mandatory.
3. Higher power plasmas, such as the 500W He-MIP, are stable during the introduction of LC effluent mist. However, LC mobile phase induced increases in the complexity and intensity of the molecular background cannot be avoided.

Future LC-MIP studies should examine nonmetal emission in the near-infrared (NIR) spectral region which is relatively free of molecular band emission caused by organic mobile phases. It may be possible, thereby, to reduce background emission effects. Other routes to be examined include the use of

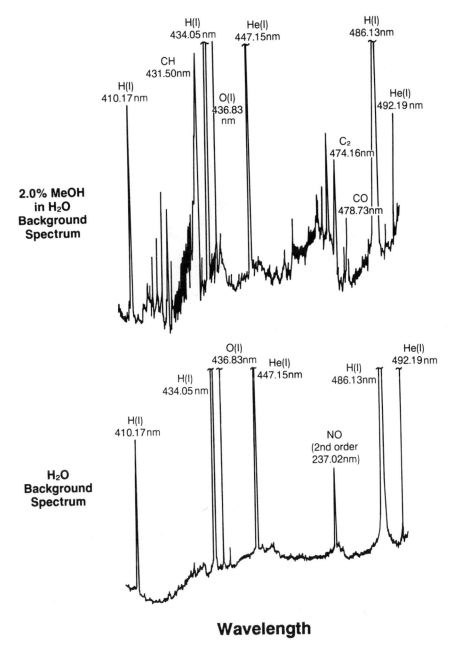

Figure 4. Spectral scans of the 400 to 500 nm range with pure water and with 2% MeOH in water with the USN-LC-MIP system. (Reproduced with permission from ref. 36.)

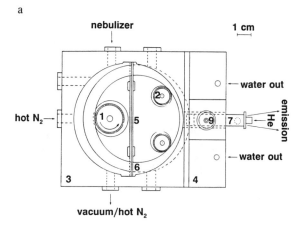

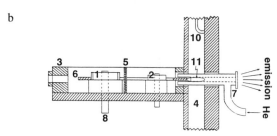

Figure 5. Schematic of the moving wheel LC-MIP interface: (a) top view; (b) side view. Key: 1. friction wheel drive; 2, guide wheel gearing; 3, cover plate; 4, TM$_{010}$ resonator; 5, separation plate; 6, stainless steel wheel; 7, fused silica plasma torch (2.5mm i.d., 7.0mm o.d.) with fused silica face plate; 8, driver wheel shaft; 9, microwave N connector; 10, coupling loop; 11, plasma region. (Reproduced with permission from ref. 14. Copyright 1989 American Chemical Society)

alternative methods to more efficiently remove the solvent prior to analyte introduction and enhanced temperature control in vaporization schemes.

Supercritical Fluid Chromatography with Helium Plasma Detection Systems

In terms of mobile phase induced perturbations on the characteristics of helium plasmas, SFC presents a situation intermediate those of GC-MIP and LC-MIP. To date, published work with supercritical fluid-helium plasma (SFC-MIP) systems describes successful routing of the mobile phase directly to the plasma. Requirements of interfaces to couple SFC with helium plasma detectors include: short distances from the column outlet (SFC restrictor) to the plasma to enhance analyte transport efficiency, heating of the interface to avoid freezing of the mobile phase expansion at the terminus of the restrictor, protection of the capillary to avoid breakage, and employment of mobile phase flows compatible with helium plasmas. Interfaces for this purpose have been designed and implemented *(15,17,37)*. Figure 6 shows such an example.

Luffer and coworkers *(15)* utilized a helium plasma sustained with a surfatron as a detector for SFC. This system was optimized for sulfur determination at 921.3 nm. Figure 7 illustrates a SFC-MIP chromatogram of thiophene derivatives. Detection limits ranged from 25 pg/s for thiophene to 150 pg/s for dibenzothiophene (S/N = 2). These detection limits are of the same order of magnitude as those of the FID and other common SFC detectors.

In the work of Skelton and coworkers *(17)*, SFC was interfaced to a radio frequency helium plasma detector (RPD), maintained with a frequency of 330 KHz. As with the 2450 MHz helium plasma, nonmetal detection is facilitated. In Figure 8, the advantages of selective detection with helium plasmas for a pesticide mixture are illustrated. Chromatogram A shows a mixture monitored using FID. Chromatograms B and C illustrate the same mixture monitored for Cl and S, respectively.

Zhang and co-workers *(37)* have used a helium MIP for SFC detection of Cl and S containing compounds. Calibration plots were linear over several orders of magnitude. Detection limits were 40 pg/s for Cl and 90 pg/s for S. A decrease in sensitivity occurred as the SFC pressure was increased.

The initial analytical data for SFC-helium plasma systems is quite promising and the implications of this mating have been characterized quite thoroughly. These studies of the effects of mobile phase flow rate upon the emission from the helium plasma are quite important because SFC separations are optimized with density/pressure programing. As the pressure of the mobile phase is increased during a programmed run, the flow rate of mobile phase effluent increases for most restrictor designs. A detailed background study was published by Galante and coworkers *(16)* using the surfatron helium plasma. Effects of SFC mobile phases upon analyte emission and spectral background with a Beenakker cavity sustained helium plasma have been studied in our laboratory *(38)*. Spectral backgrounds with SFC and the helium RPD were also monitored by Skelton and coworkers *(17)*.

As with LC, most SFC mobile phases produce intense molecular band emission in plasmas. An example of this is shown in the spectral backgrounds of a helium plasma doped with various flows of CO_2 in the ultraviolet-visible (UV-VIS) (Figure 9) and NIR spectral regions (Figure 10). Increasing amounts

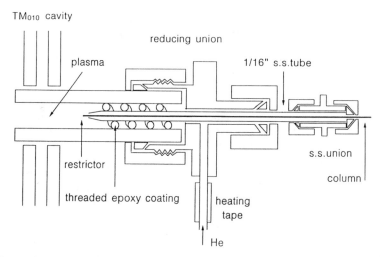

Figure 6. Schematic diagram of SFC-MIP interface. (Reproduced with permission from ref. 37.)

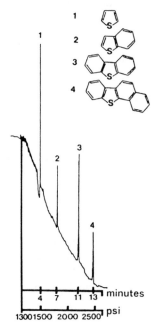

Figure 7. Separation of thiophene derivatives. Microwave power, 110W; helium flow rate, 120 mL/min; column temperature, 70°C. Approximately 60 ng of each component was introduced by a single injection. Key: 1, thiophene; 2, thionaphthene; 3, dibenzothiophene; 4, 1,2-diphenylene sulfide. (Reproduced with permission from ref. 15. Copyright 1988 American Chemical Society)

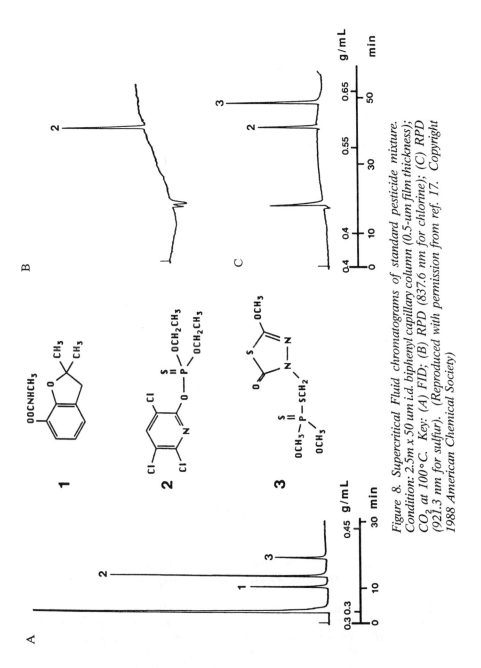

Figure 8. Supercritical Fluid chromatograms of standard pesticide mixture. Condition: 2.5 m × 50 um i.d. biphenyl capillary column (0.5-um film thickness); CO₂ at 100°C. Key: (A) FID; (B) RPD (837.6 nm for chlorine); (C) RPD (921.3 nm for sulfur). (Reproduced with permission from ref. 17. Copyright 1988 American Chemical Society)

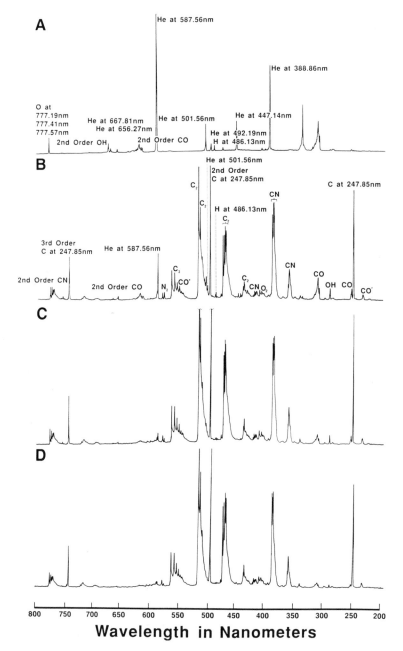

Figure 9. Background scans of CO_2 in the UV-VIS region. (A) Helium background. (B) CO_2 doping at 0.09 mL/min. (C) CO_2 doping at 0.28 mL/min. (D) CO_2 doping at 0.48 mL/min. All scans were recorded at a full scale of 100 mV, chart speed of 5 cm/min, spectral scan rate of 10 Å/s, amplifier time constant of 0.5, and a slit width of 50 microns. He flow rate for all scans was 125 mL/min. (Reproduced with permission from ref. 19. Copyright 1990 Society for Applied Spectroscopy)

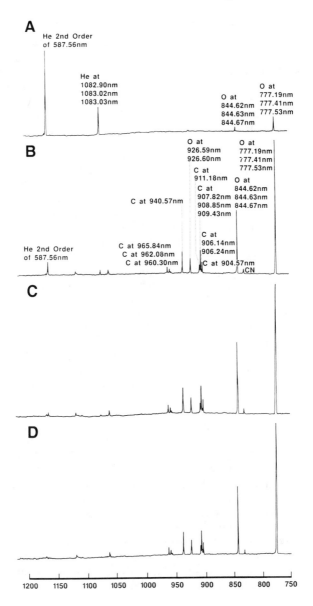

Wavelength in Nanometers

Figure 10. Background scans of CO_2 in the NIR region.   (A) Helium background.   (B) CO_2 doping at 0.09 mL/min.   (C) CO_2 doping at 0.27 mL/min.   (D) CO_2 doping at 0.52 mL/min.   All scans were recorded at the same conditions as Figure 12 except: gain, 100X; slit width, 75 microns. (Reproduced with permission from ref. 19.   Copyright 1990 Society for Applied Spectroscopy)

of CO_2 are added to the helium plasma in the lower three spectra. The UV-VIS spectral region exhibits significantly more molecular band emission from CO, C_2, CN and CO^+ than the NIR region. In the NIR spectral region, structured background emission of the plasma is predominantly atomic line emission from He and the SFC mobile phase, thus, the likelihood of spectral overlap with analyte atomic emission is much less probable. Results using N_2O are comparable, although the number of molecular emission bands is reduced *(19)*. Other researchers have reached similar conclusions *(16)*.

The addition of SFC mobile phases to helium plasmas reduce helium emission intensities. Figure 11 illustrates the effect of various gases upon helium emission at 1083.0 nm. As the flow rates of the dopant gas are increased, helium line emission intensity decreases. Figures 9 and 10 also illustrate this behavior. Both the 587.56 nm and the 1083 nm helium emission line exhibit reductions in intensity with increasing CO_2 flow. These observations indicate a change in the characteristics of high energy species in the plasma.

If the introduction of SFC mobile phases leads to reduction in the helium emission intensity, one might expect analyte emission to follow the same trend. Figures 12 and 13 illustrate the decrease of chlorine emission intensity at 479.45 and 912.11 nm for a CO_2 and N_2O doped plasmas. The monitored emission line in the UV-VIS region for chlorine arose from an ionic transition, while the emission line studied in the NIR for chlorine was from an atomic transition. These figures indicate that the intensity of the higher energy ion transition is more greatly reduced by SFC mobile phase introduction than that of the atom line. Skelton and coworkers *(17)* observed similar trends with NIR emission lines for the RPD, which showed a significant increase in detection limits with increased amounts of CO_2 using an exponential dilution system. This study was for atom lines in the NIR for sulfur and chlorine.

These background studies indicate that the NIR spectral region may be superior to the UV/VIS spectral region for SFC-MIP for many nonmetals in terms of emission signal reduction. The mobile phase composition also proved to be important. Not only does N_2O produce a different spectral background than CO_2, its effect upon emission signal intensity were less pronounced.

Ultimately, the most useful analytical emission line for each element will be determined with considerations of both signal intensity and background characteristics. Table I represents a summary of results of background studies performed in our laboratory. This table shows that each nonmetal of common analytical interest has an analytically useful emission line free of spectral interference in the presence of CO_2 and/or N_2O mobile phases.

Concerning the current status of SFC with helium plasma atomic emission detection, a few general conclusions may be included:

1. The presence of SFC mobile phases causes spectral interferences in helium plasmas, as well as decreased emission intensity of analyte signals. Both parameters are dependent upon the mobile phase flow rate and must be considered during SFC pressure programming.

2. Low power plasmas provide an adequate emission source for SFC-MIP systems. However, expected increases in mobile phase tolerances with the 500W MIP give reason to examine this plasma configuration.

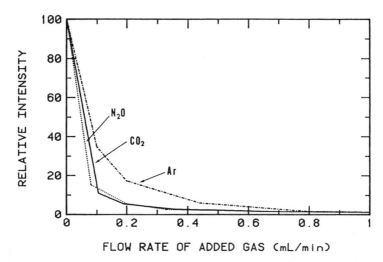

Figure 11. Effect of CO_2 (solid curve), N_2O (dotted curve), and Ar (dashed and dotted curve) on He I emission at 1083.0 nm when the gases are added to the support gas at flow rates of 0.1-1.0 mL/min (3-30 μg/s). Helium flow rate, 120 mL/min; microwave power, 100W. (Reproduced with permission from ref. 16. Copyright 1988 American Chemical Society)

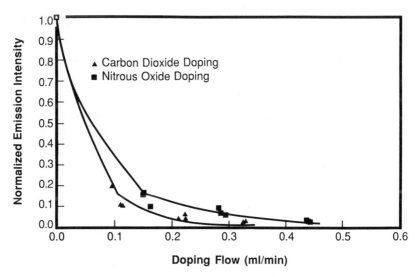

Figure 12. Cl emission intensity during CO_2 and N_2O introduction at 479.45 nm. (Adapted with permission from ref. 19. Copyright 1990 Society for Applied Spectroscopy)

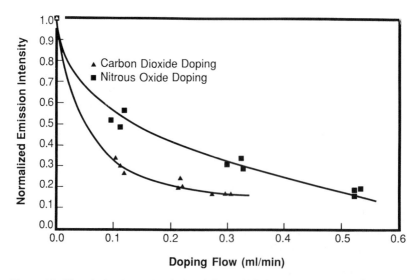

Figure 13. Cl emission intensity during CO_2 and N_2O introduction at 912.11 nm. (Adapted with permission from ref. 19. Copyright 1990 Society for Applied Spectroscopy)

Table I. Major Interferences upon Selected Nonmetal Emission Lines in the UV-VIS and NIR Spectral Regions for a N_2O and CO_2 doped He-MIP.

Element	Emission Wavelength(nm)	Possible Interference from SFC Mobile Phase	
		N_2O	CO_2
Carbon	247.86	NO band-247.9nm	mobile phase
	940.57	none observed	mobile phase
Hydrogen	486.13	N_2 band-483.7nm	none observed
	656.27, 656.28	N_2 band-654.5nm	CO band-651.4nm
Oxygen	777.19	mobile phase	mobile phase
Nitrogen	746.88	mobile phase	none observed
	868.03	mobile phase	none observed
Sulfur	545.39	none observed	C_2 band-543.5nm
	921.29	none observed	none observed
Phosphorus	213.55, 213.62	none observed	CO band-211.3nm
	979.68	none observed	none observed
Chlorine	479.45	N_2 band-472.4nm	C_2 band-473.7nm
	912.11	none observed	none observed
Bromine	470.49	N_2 band-465.0nm	C_2 band-469.8nm
	889.76	none observed	none observed
Iodine	206.24	NO band-206.2nm	CO band-206.0nm
	905.83	none observed	none observed
Fluorine	685.60	none observed	none observed
	739.87	N_2 band-738.7nm	none observed

3. The NIR spectral region seems to be superior to the UV-VIS spectral region in terms of spectral background emission characteristics.

Preliminary characterizations have born-out original justifications for the examination of helium plasmas for SFC detection. The potential of the helium plasma as a useful detector for SFC is great. Current research is addressing the problem of analyte emission reduction caused by SFC mobile phases and higher powered plasmas are being examined.

Conclusion

The potential of helium plasmas as nonmetal selective detectors for LC and SFC has driven researchers to examine the fundamental and operational characteristics of such systems. The analytical behaviors of SFC-helium plasma systems are excellent. Initial LC-helium plasma results are promising and indicate that further examinations are warranted. In the continued development of both systems, investigators must aggressively examine the relationships of mobile phase composition and introduction rate on the quality of chromatography, background spectral characteristics, the optimum applied power, and analyte signal-to-noise ratios. Successes in these areas will undoubtedly contribute to the need for additional sensitive and selective detectors for these chromatographies.

Literature Cited

1. Van Dalen, J. P. J.; de Lezenne Coulander, P. A.; de Galan, L. Anal. Chim. Acta 1977, 94, 1.
2. Brenner, K. S. J. Chromatogr. 1978, 167, 365.
3. Quimby, B. D.; Uden, P. C.; Barnes, R. M. Anal. Chem. 1978, 50, 2112.
4. Mulligan, K. J.; Caruso, J. A.; Fricke, F. L. Analyst, 1980, 105, 1060.
5. Hagen, D. F.; Belisle, J.; Johnson, J. D.; Venkateswarlu, P. Anal. Biochem. 1981, 118, 336.
6. Estes, S. A.; Uden, P. C.; Barnes, R. M. Anal. Chem. 1981, 53, 1829.
7. Slatkavitz, K. J.; Uden, P. C.; Hoey, L. D.; Barnes, R. M. J. Chromatogr. 1984, 302, 277.
8. Bradley, C.; Carnahan, J. W. Anal. Chem. 1988, 60, 858.
9. Firor, R. L. Am. Lab. 1989, 21, 40.
10. Huf, F. A.; Jansen, G. W. Spectrochim. Acta 1983, 38B, 1061.
11. Jansen, G. W.; Huf, F. A.; De Jong, H. J. Spectrochim. Acta 1985, 40B, 307.
12. Billiet, H. A. H.; van Dalen, J. P. J.; Schoemkens, P. J.; De Galan, L. Anal. Chem. 1983, 55, 847.
13. Michlewicz, K. G.; Carnahan, J. W. Anal. Lett. 1987, 20, 1193.
14. Zhang, L.; Carnahan, J. W.; Winans, R. E.; Neill, P. H. Anal. Chem. 1989, 61, 895.
15. Luffer, D. R.; Galante, L. J.; David, P. A.; Novotny, M.; Hieftje, G. M. Anal. Chem. 1988, 60, 1365.
16. Galante, L. J.; Selby, M.; Luffer, D. R.; Hieftje, G. M.; Novotny, M. Anal. Chem. 1988, 60, 1370.

17. Skelton, R. J.; Farnsworth, P. B.; Markides, K. E.; Lee, M. L. Anal. Chem. **1989**, 61, 1815.
18. George, M. A.; Hessler, J. P.; Carnahan, J. W. J. Anal. Atom. Spectrom. **1989**, 4, 51.
19. Webster, G. K.; Carnahan, J. W. Appl. Spectrosc. **1990**, 44, 1020.
20. Michlewicz, K. G.; Urh, J. J.; Carnahan, J. W. Spectrochim. Acta **1985**, 40B, 493.
21. Michlewicz, K. G.; Carnahan, J. W. Anal. Chem. **1985**, 57, 1092.
22. Michlewicz, K. G.; Carnahan, J. W. Anal. Chim. Acta **1986**, 186, 275.
23. Michlewicz, K. G.; Carnahan, J. W. Anal. Chem. **1986**, 58, 3122.
24. Mazzo, D. J.; Elliott, W. G.; Uden, P. C.; Barnes, R. M. Appl. Spectrosc. **1984**, 38, 585.
25. Uden, P. C.; Quimby, B. R.; Barnes, R. M.; Elliott, W. C. Anal. Chim. Acta **1978**, 101, 99.
26. Nisamaneepong, W.; Caruso, J. A.; Ng, K. C. J. Chromatogr. Sci. **1985**, 23, 465.
27. Yoshida, K.; Haraguchi, H. Anal. Chem. **1984**, 56, 2580.
28. Nisamaneepong, W.: Ibrahim, M.; Gilbert, T. W.; Caruso, J. A. J. Chromatogr. Sci. **1984**, 22, 473.
29. Ibrahim, M.; Nisamaneepong, W.; Caruso, J. A. J. Chromatogr. Sci. **1985**, 23, 144.
30. Ibrahim, M.; Gilbert, T. W.; Caruso, J. A. J. Chromatogr. Sci. **1984**, 22, 111.
31. Morita, M.; Uehiro, T.; Fuwa, K. Anal. Chem. **1980**, 52, 351.
32. Hausler, D. W.; Taylor, L. T. Anal. Chem. **1981**, 53, 1223.
33. Savage, R. N.; Smith, S. B. J. Liq. Chromatogr. **1982**, 5, 463.
34. Fraley, D. M.; Yates, D.; Manahan, S. E. Anal. Chem. **1979**, 51, 2225.
35. Allen, L.; Koropchak, J. A. Meeting of the Federation of Analytical Chemistry and Spectroscopy Societies, Chicago, IL, 1989, Abstract #214.
36. Michlewicz, K. M. Ph.D. Dissertation, Northern Illinois University, 1987.
37. Zhang, L.; Ph.D. Dissertation, Northern Illinois University, 1990.

RECEIVED April 12, 1991

Chapter 3

Quantitative Characteristics of Gas Chromatography with Microwave-Induced Plasma Detection

Weile Yu, Yieru Huang, and Qingyu Ou

Lanzhou Institute of Chemical Physics, Chinese Academy of Sciences, P.O. Box 97, Lanzhou 730000, People's Republic of China

A survey of previous and original data on quantitative elemental responses and empirical formula determination is presented for GC with atmospheric pressure MIP detection. The accuracy of ratios and empirical formulae obtained is found to depend on the closeness of molecular structure and elemental composition of the reference compounds chosen in the determination of unknowns. The determination of hydrocarbons is found to be less dependent on this molecular similarity than is that of chlorinated and brominated compounds.

For the chromatographer, separation is not the final definitive procedure, since the monitoring and identification of eluted chromatographic peaks is vital to the effectiveness of separation. Detection may provide various modes of selectivity, as exemplified by the ECD, which is strictly speaking a group specific detector that is very sensitive to halogens and some electrophores. It is not an element specific detector, since it cannot differentiate individual halogen elements from each other. Furthermore, the sensitivity and selectivity of the ECD is markedly affected by the structure or elemental composition and electronic cross-section of a given compound. The ECD is very insensitive to CH_3Cl and CH_2Cl_2 (1), and response for Freon-11 (CCl_3F) is 70 times larger than that for Freon-21 ($CHFCl_2$) (2). The only elemental difference between the Freons is that the former has three chlorine atoms and the latter two.

Since McCormack et al. (3) described the characteristics of a microwave induced plasma (MIP) detector for GC, many have been attracted by its potentialities as a selective, universal, and potentially quantitative detector, with features to perhaps surpass conventional detectors. Dagnall et al. (4) were the first group to use GC-MIP to measure the interelemental ratios of chlorine to carbon, sulfur to carbon, bromine to

0097–6156/92/0479–0044$06.00/0

carbon, iodine to carbon, and phosphorus to carbon, obtaining maximum deviations from the theoretical ratio of 5-8 percent.

Reviews on empirical and molecular formula determination by GC-MIP have been prepared by Uden et al. (5,6) and Caruso et al. (7,8), and efforts to improve the accuracy and precision of empirical formula determination are continuing (9-12). A laminar flow torch described by Bruce and Caruso (10) showed very encouraging results with respect to cost, simplicity, long-term stability, and low helium flow rate. Evans et al. (9) used a tangential-flow version of the 'Beenakker' cavity at 20 torr, finding that hydrogen/carbon ratios for sulfur compounds were somewhat less precise than those for the pure hydrocarbons. They noted that this design combined some of the best features of the atmospheric-pressure Beenakker and the Evenson reduced-pressure sources. Sklarew et al. (11) determined empirical formulae of several thiophenes with the same system, obtaining good agreement of calculated and true values for carbon and fair correspondence for hydrogen. Moussounda et al. (12) described various discharge configurations of atmospheric pressure microwave plasmas which could influence the completeness of fragmentation of the compound entering the discharge tube. Freeman et al. (13) and Pivonka et al. (14) suggested using near- infrared and Fourier transformed far red/near-infrared atomic emissions respectively, instead of the conventional emission lines in the UV-visible region, to determine elemental ratios of C, H, F, Cl, Br, S, N and O. Haas et al. (15) and Hagen et al. (16) emphasized the importance of background correction of the carbon continuum, which otherwise significantly degraded the accuracy of the elemental ratios.

The appropriate selection of reference compounds is another important factor affecting the accuracy of empirical formula determination. Hagen et al. (16) and Slatkavitz et al.(17) pointed out that reference compounds should be chosen to elute as closely as possible to the analyte peak. Hagen et al. (16) also found it necessary to select a structurally similar reference to the unknown peak in order to obtain more credible results. The dual homolog derivatization method (16) and the multi-referencing technique (18) were both intended to provide ideal references for quantitation purpose. Dagnall et al. (19) mentioned that a platinum catalyst could be used to fragment stable compounds, and the detection limits for thiophene, dimethyl sulphoxide, and glycollic acid were improved by using a platinum loop. The reason for this improvement was considered to be that platinum would partially vaporize in the discharge and aid in the fragmentation process. Chiba and Haraguchi (20) lengthened the discharge tube from 10 cm to 20 cm and heated it to 900°C aiming to achieve complete pyrolysis of eluted compounds. Dingjan and deJong (21) also commented that a thermochemical "cracker" or pyrolysis unit in front of the "cool" MIP could diminish intercompound differences. Uden et al. (5) reported that it was unnecessary to use a pyrolyzer for compounds with carbon number of less than 24, but based on their experience with the empirical formula determination of cyclosiloxane pyrolyzates (6), the use of a pyrolyzer in front of the microwave discharge tube to overcome "incomplete fragmentation" of eluted compounds was deemed useful.

In order to evaluate the quantitative behavior of GC- MIP systems, relative response measurements or relative sensitivities and empirical formula determinations have often been made (22-27). For atmospheric pressure systems, relative halogen responses for chlorine (versus $CHCl_3$) ranged from 0.95 for CCl_4 to 1.01 for $CHBrCl_2$, for bromine (versus $CHBr_3$) from 1.03 for $CHBr_2Cl$ to 1.07 for $CHBrCl_2$, and for iodine (versus CH_3I) from 0.96 for $CH_3CH_2CH_2I$ to 1.06 for CH_3CH_2I (23). In another study, relative chlorine response ranged from 0.97 for 1-chlorohexane to 0.55 for DDT (24). Carbon response (versus $CHCl_3$) ranged from 1.07 for CCl_4 to 0.73 for thiophene (25), and from 1.11 for CCl_4 to 0.48 for methanol (versus n-hexane) (26). For reduced pressure systems, carbon response ranged from 0.75 for chlorobenzene to 0.93 for $CHCl_3$ (versus CH_2Cl_2) (21). Bromine response ranged from 1.05 for 1-bromopentane to 1.26 for dibromopentane (versus. bromobenzene), and chlorine response from 1.05 for tetrachloroethane to 1.25 for $CCl4$ (versus chlorobenzene) (27). Hydrogen, oxygen and sulfur showed a similar or greater response range in the latter study.

Computed empirical formulae for 'CH$_2$' ratio hydrocarbons ranged from $CH_{1.652}$ for cyclohexene to $CH_{2.052}$ for 1-hexene (versus ethylbenzene) (28). Values (versus hexane) were from $C_4H_{9.5}$ for butane to $C_7H_{16.6}$ for heptane (29), and from $C_{11}H_{21.9}$ for undecene to $C_{12}H_{26.4}$ for dodecane (versus hexadecane) (30). Other studies showed similar data ranges for CH ratios (5,9). For halogens, typical CCl ratios were $CCl_{3.69}$ for CCl_4, $C_6Cl_{1.06}$ for chlorobenzene (versus CH_2Cl_2) (21). CHCl ratios were $CH_{0.94}Cl_{2.84}$ for $CHCl_3$ and $CH_{2.9}Cl_{1.2}$ for CH_3Cl (versus 1,1,2-trichloroethane) (31), and CCl ratios were $C_4Cl_{6.3}$ for hexchlorobutadiene and $C_{12}Cl_{1.2}$ for 1-chlorodecane (versus 1,2,4-trichlorobenzene) (32). Typical ratios for oxygenated compounds were $C_3H_{5.9}O_{0.92}$ for n-propanol and $C_8H_{18}O_{1.3}$ for dibutyl ether (versus dioxane) (31). Values for deuterium were typified by $C_{23}H_{27.56}D_{4.55}O_{2.29}$ for pentadeutero-oestradiol cyclopentyl ether ($C_{23}H_{28}D_5O_2$) (33) and $C_3D_{5.74}O_{0.99}$ for perdeuteroacetone (Yieru Huang, Lanzhou Institute of Chemical Physics, unpublished data). Values for nitrogen were typically within 5% of true values (34-36).

The deviations of some data may be caused by experimental error, and the importance of minimizing the empirical formula determination error from spectroscopic sources (22) (such as microwave input power, plasma flow rate, background interference, spatial viewing position, plasma length, wall effect of discharge tube, wavelength used etc.) and chromatographic sources (such as column resolution and stability, purity of carrier gas, etc.) parameters has been generally agreed. The influence of background noise and the choice of calculation method need also to be considered. However, the data surveyed indicates that the chemical structures of compounds may sometimes affect empirical formula determination and the relative element response.

This paper surveys previous and new results of the evaluation of a prototype Model SG-2 GC- atmospheric pressure MIP atomic emission spectrometry system for the relative elemental response and empirical formula determination of carbon, hydrogen, chlorine, bromine, and oxygen- containing compounds with various chemical structures (31, 37-39, Yieru Huang, Lanzhou Institute of Chemical Physics, unpublished data). Factors affecting accuracy and precision are discussed.

Experimental Section

Instrumentation. The specifications of the prototype Model SG-2 GC-MIP instrument were as follows: Microwave generator (Huohua Glass Instrument Works, Nanjing, PRC) with output power 0-200W at 2450±30MHz. Beenakker TM010 Cavity (J and D. Instrument Co., Lexington, MA, USA). Quartz Plasma Discharge tube, length 7.5 cm, 6.5 mm o.d., 0.6 mm i.d.; Spectrometer (The 2nd Beijing Optical Instrument Works, Beijing, PRC), wavelength range 400-800 nm (1st order), 200-400 nm (2nd order); radius of curvature 750 mm; grating, 1200 grooves/mm, blazed at 550 nm, and reciprocal linear dispersion of 1.1 nm/mm; entrance slit 25 mm, exit slit, 50 mm or 75 mm.; Photomultiplier tubes, Hamamatsu 1P28, R300, R427 and R446; Gas Chromatograph, Model 1001 (Shanghai Analytical Instrument Works, Shanghai, PRC); GC Columns: i) SE 52 Fused Silica Capillary, (FSOT), 30 m x 0.315 mm i.d, 0.25 mm film thickness (J and W Scientific, Folsom, CA) for determinations of hydrocarbons and brominated compounds; ii) SE 54 wide bore, thick film glass capillary column, 30 m x 0.48 mm i.d. (prepared in the laboratory) for chlorinated compounds; iii) DB 5 FSOT, 30 m x 0.32 mm i.d. (J and W Scientific).

Reagents. Purities of the chemicals used were: n-paraffins, n- olefins, alcohols- chromatographically pure; aromatic and polynuclear aromatic hydrocarbons, oxygenated compounds- analytically pure; chlorinated and brominated compounds - analytically or chemically pure (purity > 98%).

Procedure. Plasma operating conditions were 65W and 70W input power for hydrocarbons and halocarbons respectively, and 45 ml/min helium flow rate. Chromatographic conditions varied for different samples. The emission lines used were C(I), 247.86nm (2nd order), H(I) 486.13nm, Cl(II) 479.5nm, Br(II) 470.49nm, where I indicates an atomic emission line and II an ionic emission line.

Relative elemental responses, per mole atom, were determined by dividing the peak area by the ratio of element amount (g) to atomic weight. Empirical formulae were determined for various carbon, hydrogen, chlorine and bromine- containing compounds. The responses for simultaneous multi-element detection were recorded by stripchart recorder and peak areas were measured manually by planimeter. Specific element- to- carbon ratios were determined as follows: the elemental ratio (unknown) equals the elemental ratio (reference), times the elemental signal ratio (unknown), divided by the elemental signal ratio (reference) In each chromatogram, one compound was designated as the "reference" and all others were treated as "unknown".

Results and Discussion.

Relative Response Determination.

Hydrocarbons. Representative paraffinic, olefinic, and aromatic hydrocarbons showed no significant differences in relative responses per mole of carbon and hydrogen, regardless of structure. Using n-undecene as a reference, with elemental responses per mole atom defined as 1.00, alkanes and alkenes showed similar results at 1.00± 0.05; polynuclear aromatics' responses were somewhat lower, but this had no significant effect on quantitative analysis, so no calibration was needed for this type of mixture. Relative responses were also unaffected by change of microwave energy from 60 to 80W (39).

Mixtures of olefinic hydrocarbons, aromatic hydrocarbons, and chlorinated compounds. Relative responses of carbon and hydrogen for different hydrocarbons showed no serious deviation, but for chlorinated compounds the relative responses of carbon and hydrogen increased notably with increase in the number of chlorine atoms in the compound (Table I). The relative responses of hydrogen increased more than those of carbon.

Brominated compounds. Table II shows that the relative responses of hydrogen increased more significantly with an increase in the bromine atom number of the compound measured. The relative responses of carbon, hydrogen and bromine for brominated aromatics (bromobenzene, bromomethylbenzene, dibromobenzene) were higher than those for the brominated paraffin hydrocarbons with the same carbon atom number. Changing the microwave power from 60-80W had no effect on the relative responses of carbon, hydrogen and bromine.

Oxygenated compounds. No noticeable effect was found, due to the presence of oxygen, on the relative responses of carbon and hydrogen for the compounds listed in Table III. The use of different microwave energies from 70-90W again did not change the response factors.

It was considered that chlorine and bromine emission could not interfere spectroscopically with the carbon and hydrogen emission lines since there were no strong halogen emissions nearby. Changing the viewing position for the halogens produced no effect on the relative responses of carbon or hydrogen for the chlorinated compounds. Freeman and Hieftje (40) noted that the excitation intensity and the temperature of the helium plasma decreased in the presence of a hydrocarbon. This could explain the increase in the relative responses of carbon and hydrogen with compounds with higher halogen atom numbers since the carbon and hydrogen content is comparatively low. They also mentioned that, for different elements, excitation temperatures might be expected to vary according to the ionization potential of the element. We believe that fragmentation and excitation efficiencies may depend on the structures of these sample molecules as mentioned by Freeman and Hieftje (13).

Table I. Relative Elemental Responses per Mole Atom

Compound	Molecular Formula	Carbon			Hydrogen			Halogen		
		60W	70W	80W	60W	70W	80W	60W	70W	80W
Ethylbenzene	C_8H_{10}	1.01	1.02	0.96	0.99	0.96	1.00			
m-Dimethylbenzene	C_8H_{10}	1.07	0.96	1.05	1.09	0.97	1.11			
1,3,5-trimethylbenzene	C_9H_{12}	0.95	0.99	0.94	0.98	1.01	1.00			
Decene	$C_{10}H_{20}$	1.04	1.00	0.99	1.01	1.01	1.00			
Undecene	$C_{11}H_{22}$	1.00	1.00	1.00	1.00	1.00	1.00			
Hexadecene	$C_{16}H_{32}$	1.03	1.00	0.96	1.02	0.99	0.96			
Chlorododecane	$C_{12}H_{25}Cl$	0.89	0.94	0.90	0.94	0.95	0.97	0.68	0.82	0.63
m-Dichlorobenzene	$C_6H_4Cl_2$	1.01	1.06	1.07	1.23	1.29	1.26	1.00	1.00	1.00
1,2,3-Trichloropropane	$C_3H_5Cl_3$	1.20	1.18	1.16	1.31	1.34	1.32	1.07	1.05	1.02
1,1,2,2-Tetrachloroethane	$C_2H_2Cl_4$	1.56	1.44	1.51	1.70	1.76	1.67	1.02	1.04	1.03
Pentachloroethane	C_2HCl_5	1.96	1.84	1.82	3.03	2.78	2.44	1.01	1.02	1.05

Table II Relative Elemental Responses per Mole Atom of Brominated Hydrocarbons

Compound	Molecular Formula	Carbon			Hydrogen			Halogen		
		60W	70W	80W	60W	70W	80W	60W	70W	80W
Decene	$C_{10}H_{20}$	1.00	1.00	1.00	1.00	1.00	1.00			
Bromopropane	C_3H_7Br	1.01	1.05	1.03	1.05	1.04	1.03	1.00	1.03	1.01
Bromobenzene	C_6H_5Br	1.08	1.14	1.14	1.21	1.18	1.21	1.07	1.09	1.06
p-Bromomethyl-benzene	C_7H_7Br	1.13	1.12	1.12	1.10	1.14	1.14	1.17	1.13	1.22
Bromodecane	$C_{10}H_{21}Br$	1.04	1.04	1.03	1.03	1.03	1.03	0.98	0.96	0.96
Dibromomethane	CH_2Br_2	1.16	1.12	1.14	1.34	1.26	1.34	1.06	0.98	1.02
1,3-Dibromopropane	$C_3H_6Br_2$	1.08	1.10	1.08	1.15	1.17	1.15	1.12	1.18	
Bromoform	$CHBr_3$	1.28	1.22	1.20	2.00	2.01	1.96	1.17	1.21	

Empirical Formula Determination.

Hydrocarbons. Mixtures of C_{10} to C_{18} normal alkenes (excluding pentadecene and heptadecene), a mixture of mononuclear aromatic hydrocarbons, and a mixture of polynuclear aromatic hydrocarbons were studied. In addition, to investigate the determination of empirical formulae when using non-homologs as reference compounds, a mixture of normal alkanes and alkenes, and a mixture of polynuclear aromatic hydrocarbons and alkenes were chosen as test samples.

Theoretical molecular formulae and experimental empirical formulae for normal alkenes were determined. The relative standard deviation for formula determinations of seven alkenes, based on three determinations was less than 3.5%. Relative errors between the theoretical and experimental formulae were 1-2% . When each compound in this mixture was chosen sequentially as "reference", similar accuracy was obtained; the difference in elution rates between the reference and the "unknown" had no significant effect on the determination errors.

The relative standard deviations determined of formulae for six mononuclear aromatic hydrocarbons, based on four determinations, was less than 4% and the relative error was less than 6%. The relative standard deviation for polynuclear aromatic hydrocarbons ranged from 0.6 to 8%, and the relative error between theoretical and experimental formulae was 0.5 to 6%. Aromatic hydrocarbons with comparatively fewer hydrogens than those of paraffins had lower hydrogen signals, therefore small variations in measurement greatly influenced the accuracy of elemental response calculation. In the case of homologous hydrocarbons, the chemical structures of compounds had no significant effect on the elemental response.

The relative standard deviations of formulae for the mixture of alkanes and alkenes was 0.2 - 4%, and the relative error between theoretical and experimental formulae of each compound was less than 2%. Whenever an alkane or alkene was chosen as the reference compound, the experimental and theoretical formulae agreed reasonably well. The relative standard deviations for the mixture of polynuclear aromatic hydrocarbons and alkenes were from 0.2 - 4 %, based on five determinations, and the relative error of each compound was 0.1 - 8%. When alkenes were used as references, the deviations for alkenes and polynuclear aromatic hydrocarbons from the theoretical values were less than 2.3 and 8%, respectively. In contrast, when polynuclear aromatic hydrocarbons were used as references for the determination of the polynuclear aromatic hydrocarbons-alkenes mixture, the relative error between the empirical formulae of alkenes and the theoretical values increased, but it decreased for polynuclear aromatics . In order to improve the accuracy of formula determination, it was clearly necessary to choose a reference compound having a similar structure to that of the analyte compounds.

Brominated hydrocarbons. The error between experimental and theoretical data for the number of hydrogen atoms in the brominated compounds varied little when using different references as is shown in Figure 1. For tetrabromoethane as reference, the empirical formulae errors were at least three times greater than for other reference

Table III. Relative Elemental Responses per Mole Atom of Oxygen-containing Compounds

Compound	Molecular Formula	Carbon			Hydrogen		
		60W	70W	80W	60W	70W	80W
Dodecene	$C_{12}H_{24}$	1.00	1.00	1.00	1.00	1.00	1.00
Heptanol	$C_7H_{16}O$	1.06	1.01	0.98	1.01	0.94	0.96
Nonanol	$C_9H_{20}O$	0.90	0.96	0.90	0.86	0.95	0.90
Ethyl heptanoate	$C_9H_{18}O_2$	0.99	1.12	1.17	1.15	1.15	1.16
Methyl undecanoate	$C_{12}H_{24}O_2$	0.90	0.98	0.94	0.94	1.02	0.98
Methyl dodecanoate	$C_{13}H_{26}O_2$	0.88	0.91	0.87	0.91	0.96	0.87

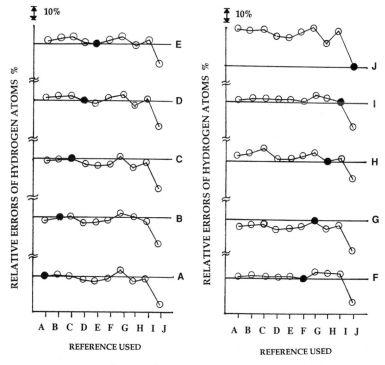

Figure 1. Plots for the relative errors for hydrogen atoms in the empirical formulae of brominated hydrocarbons determined vs reference used.
A) n-bromopropane, B) n-bromobutane, C) bromobenzene, D) n-bromoheptane,
E) n-bromooctane, F) dibromomethane,G) 1,2-dibromopropane,
H) 1,3-dibromopropane, I) dibromobenzene, J) 1,1,2,2-tetrabromoethane.

compounds. Good accuracy was obtained for the determination of mono-brominated hydrocarbons when other mono-brominated hydrocarbons were used as references, the largest error being only 0.1 bromine atoms. The determination of polybrominated hydrocarbons was also accurate when other polybrominated hydrocarbons were used as references, with maximum errors of only 0.2 and 0.3 bromine atoms for dibromo- and tetrabromo- hydrocarbons respectively (Figure 2). These results again demonstrated the close relationship between accuracy of empirical formula determination and compound structure.

Chlorinated hydrocarbons. The relative error between empirical and theoretical formulae for the number of hydrogen atoms varied slightly with structure of compounds as summarized in Figure 3. The maximum error of determination for all compounds, except pentachloroethane and tetrachloroethane, was approximately 10%. The error in hydrogen to carbon ratios was greater as the number of chlorine atoms increased. Using a monochlorohydrocarbon as the reference, the error of empirical formulae determination of other monochlorohydrocarbons was 0.1 chlorine atoms, but errors for dichloro-, trichloro-, tetrachloro-, and pentachloro-hydrocarbons were 0.4, 0.7, 0.8, and 1.1 chlorine atoms respectively. If a polychloro-hydrocarbon was used as reference, the error for monochloro-compounds increased to 0.3 chlorine atoms and that for dichloro-, trichloro-, tetrachloro-, and pentachloro- hydrocarbons decreased correspondingly to 0.2, 0.4, 0.4, and 0.5 chlorine atoms respectively (Figure 4).

There was clearly a correlation between the accuracy of empirical formula determination, the reference used, and the compound structure of the analyte. It is well known that when carbon-containing compounds are eluted into the plasma, they contribute a potential source of determination error in multielement detection. The background intensity shifts at the monitored specific element line due to emission from molecular species such as CN, CO, and C_2+, resulting in the deviation of heteroelement to carbon ratios. Background correction techniques are thus vital.

The analytical emission lines of chlorine and bromine used in this study were ion lines in the UV-visible region having higher excitation energies than atomic emission lines. If chlorine and bromine atoms were not fully ionized in the plasma, leaving a mixture of ions and atoms, then the elemental responses detected for ionic emission would deviate. Because of the different number of halogen atoms present in different compounds, the atom/ion ratios of these compounds in the plasma may differ, giving rise to the deviation in results.

Dingjan and deJong (21) mentioned that high molecular weight compounds tend to be incompletely fragmented. We observed that the relative responses of halogen to hydrogen atoms increased with the number of halogen atoms, this behavior countering the idea of incomplete fragmentation of large molecular weight compounds. Our data suggests that the errors in empirical formula determinations result mainly from the number of heteroatoms and the molecular structure of the anticipated analyte. In the case of halogenated compounds, a reference of similar structure, number of

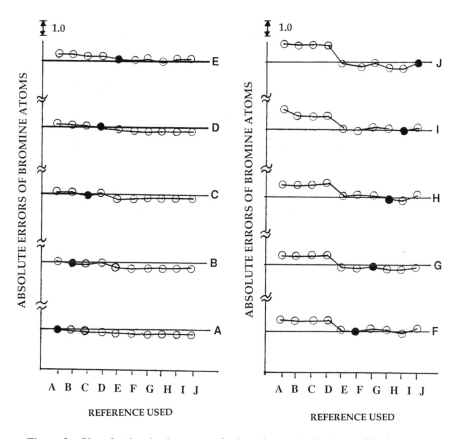

Figure 2. Plots for the absolute errors for bromine atoms in the empirical formulae of brominated hydrocarbons determined vs reference used.A), B) and F-J) as in Figure 1.C) n-bromoheptane, D) n-bromooctane, E) bromobenzene.

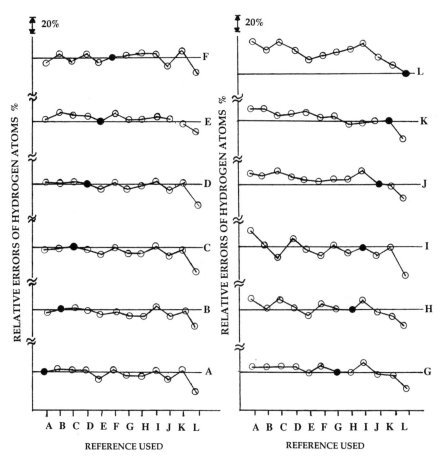

Figure 3. Plots for the relative errors of hydrogen atoms in the empirical formulae of chlorinated hydrocarbons determined vs reference used. A) n-chloropropane, B) n-chloropentane, C) chlorocyclohexane, D) chlorododecane, E) α-chloronaphthalene, F) dichloromethane, G) 1,2-dichloroethane, H) m-dichlorobenzene, I) chloroform, J) 1,1,2-trichloroethane, K) 1,2,3-trichloropropane, L) 1,1,2,2-tetrachloroethane,

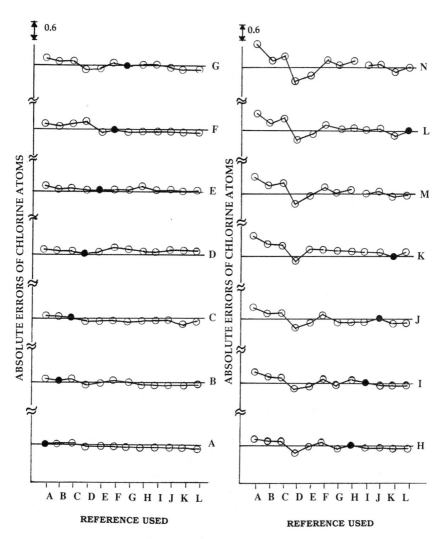

Figure 4. Plots for absolute errors for chlorine atoms in the empirical formulae of chlorinated hydrocarbons determined vs reference used. A) - L) as Figure 4. M) tetrachloroethene, N) pentachloroethane.

heteroatoms, and retention time to the eluted peak improves the accuracy of the empirical formulae obtained.

Applications

An analysis of impurities in Freon- 12 (41) (Figure 5) illustrates clearly that the elemental ratios of carbon/fluorine and carbon/chlorine for peaks 7 and 8 are different because of the one chlorine atom difference. This elemental ratio difference aids confirmation of molecular formulae and provides reliable identification of each eluted peak. As mentioned earlier, the ECD response for Freon 21 is anomalous and heretofore the only reliable analytical method has been by high resolution mass spectrometry. With GC- MIP, Freon 11 and Freon 21 can be analyzed quantitatively and qualitatively (see Tables IV-V).

Another quantitative application of GC - MIP is the examination of purity of eluted peaks using the following calculation method (38), i.e.,

Normalization method:

$$W_i \% = \frac{A_i M_i / C_i}{\sum A_i M_i / C_i} \times 100 \qquad \text{(Eqn 1)}$$

Internal reference method:

$$W_i \% = \frac{A_i m_s M_i / C_i}{A_s m_o M_s / C_s} \times 100 \qquad \text{(Eqn 2)}$$

in which W_i = weight % of the compound determined; A_i = peak area of the compound determined for the specific element channel; A_s = peak area of the internal reference for specific element channel; M_i = molecular weight (calculated from empirical formula) of the compound determined; C_i = total atomic weight of the specific element in the compound determined; C_s = total atomic weight of the specific element in the internal reference; m_s = weight of the internal reference; m_o = weight of the sample used, and M_s = molecular weight of the internal reference.

The quantitative content of acetone, t-butanol, and ethyl acetate in a mixture could be calculated with reference to carbon or oxygen by either equation 1 or 2, using the GC- MIP responses for the carbon or oxygen channels. The proportions of these three compounds were calculated as 1.52:1:1.6 respectively with reference to carbon, and as 1.40:1:0.86 with reference to oxygen. The significant reduction of ethyl acetate content when calculated with reference to oxygen indicated that this peak might be a mixture and not pure ethyl acetate. The original ethyl acetate peak could be separated into two peaks by employing a capillary column of different polarity, one peak being ethyl acetate and the other hexane. This example also demonstrated that even for a peak containing more than one component, one can still calculate the quantity of the component having the specific heteroatom which differs from that of other components, by using the response of the specific heteroatom channel. For example,

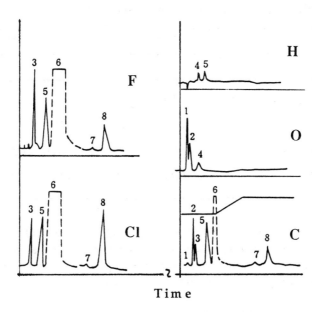

Figure 5. Multi-element detection of Freons. Column 3.7 m x 2mm i.d. Chromosorb 102. Peak identification according to Table IV.

Table IV. Qualitative Analysis of Impurities in Freon-12

-Peak No.	Empirical Formula Determined	Corresponding Compound
1	O H	air
2	$CO_{2.01}$	CO_2
3	$CF_{2.95}Cl_{1.04}$	CF_3Cl (Freon-13)
4	O H	H_2O
5	$CH_{1.04}F_{1.95}Cl_{1.04}$	CHF_2Cl (Freon-22)
6	internal standard	CHF_2Cl (Freon-12)
7	$CH_{0.96}F_{0.98}Cl_{2.05}$	$CHFCl_2$ (Freon-21)
8	$CF_{1.01}Cl_{3.08}$	$CFCl_3$ (Freon-11)

Table V. Quantitative Analysis of Impurities in Freon-12

| | | Mole Percent | | | | | |
| | | Sample 1 | | | Sample 2 | | |
		C	F	Cl	C	F	Cl
CF$_3$Cl	Freon-13	0.046	0.037	0.041			
CHF$_2$Cl	Freon-22	0.28	0.23	0.26	0.15	0.14	0.14
CHFCl$_2$	Freon-21				0.047	0.046	0.048
CFCl$_3$	Freon-11	0.12	0.12	0.13	0.16	0.17	0.17

when oxygenated compounds (31) or alkyl leads (42) are mixed with hydrocarbons, their complete separation from them is not required since they can be determined quantitatively and selectively through the oxygen or lead channels respectively. Thus the entire analytical procedure is simplified considerably.

Conclusions and Projections.

The structures of paraffin, olefin, and aromatic hydrocarbons and the presence of oxygen appear to have little or no effect on the relative elemental responses of carbon and hydrogen. For polynuclear aromatic hydrocarbons, the relative responses of carbon and hydrogen are only slightly less than for paraffins and olefins, so no calibration is required. For halogenated compounds, relative responses for carbon and hydrogen increase with an increase in the number of halogen atoms in the molecule. The relative responses of chlorine do not vary with the molecular structure in polychlorinated compounds, but responses for bromine increase with an increase in the bromine atom number. Those compounds having the same bromine atom number have similar elemental responses. Microwave power variation between 60 and 90W has no significant influence on the mole atom response of carbon, hydrogen, chlorine, or bromine. For a multi-component mixture containing polyhalogenated compounds, it is advisable to use a calibration factor, or to select a reference compound having both similar structure and similar halogen atom number.

For homologous hydrocarbons, accuracy is not affected by the reference adopted, so any compound in the homolog series can be used as the reference. For a mixture of alkenes and alkanes, the results of empirical formulae determination do not generally vary with the reference. In the case of non-homologs, such as mixture of alkenes and polynuclear aromatic hydrocarbons, the structure of the reference compound affected the accuracy of empirical formulae to some extent. For example, for an alkene reference, the results of alkenes in this mixture are more accurate than those obtained with a polynuclear aromatic hydrocarbon reference. Likewise, for a

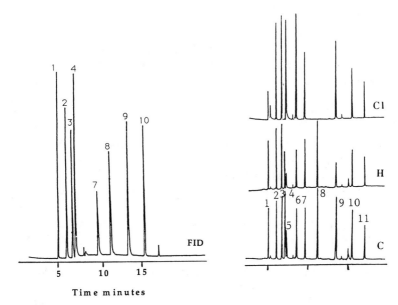

Figure 6. Comparison GC/FID and MIP chromatograms. Column 50m x 0.28 mm i.d. cross-linked glass capillary column. FID at 200°C, MIP 70W, components ca. 1% in dodecane. 1) tert-butyl chloride, 2) chloroisobutane, 3) chlorobutane, 4) benzene, 5) chloroacetonitrile, 6) 2-chloro-1,3-epoxypropane, 7) chloro-n-pentane, 8) n-octane, 9) chlorobenzene, 10) chlorocyclohexane, 11) 1,1-diethoxy-2-chloroethane.

polynuclear aromatic hydrocarbon reference, better results are obtained for empirical formulae for other polynuclear aromatics. For the halogenated hydrocarbons, the chosen reference compound should have a similar carbon-hydrogen skeleton to the tested compounds and a similar number of atoms of the halogen. From consideration of the chromatographic retention behavior of analyte compounds, molecular formulas can be readily deduced from the empirical formulas determined by GC- MIP.

In spite of some interelemental interferences and molecular structure effects, GC- MIP has the following attributes:

1. It provides a universal and selective multi-element system for organic and inorganic analysis. For example,it can differentiate between halogens and detect oxygenated and deuterated compounds.

2. It can provide useful supplementary to low resolution mass spectrometry and can give better elemental information, as shown by the cited Freon analysis.

3. It is a relatively inexpensive detector that can be used for a preliminary rapid scan of a complex mixture to indicate elemental components.

4. The low detection limits for most elements, comparable to or better than other sophisticated GC detectors, ensure that few elutable compounds are undetected, as exemplified in Figure 6 in which peaks 5 and 6 are undetected and peak 11 is poorly detected by FID.

Acknowledgements.

The authors are grateful to Peter C. Uden for his helpful discussions and contribution to the final version of this paper. Yadi Zeng is also thanked for her help in compiling the paper.

Literature Cited.

(1)   Cronn, D. R.; Harsch, D. E. *Anal. Letters*, **1979**, *12(B14)*, 1489.
(2)   Crescentini, G ; Mangani E.; Mastrogiacomo A. R.; Bruner, F. *J. Chromatogr.*, **1981**, *204*, 445.
(3)   McCormack, A. J.; Tong, S. G.; Cooke W. D. *Anal. Chem.*,**1965**, *37*, 1470.
(4)   Dagnall, R. M.; West , T. S.; Whitehead, P. *Anal. Chem.*, **1972**, *12*, 2074.
(5)   Uden, P. C.; Slatkavitz, K. J.; Barnes, R. M. *Anal. Chim. Acta,* **1986**, *180*, 401.
(6)   Uden, P. C.; Yoo, Y.; Wang, T.; Cheng, Z. B. *J. Chromatogr.*, **1989**, *468*, 319.
(7)   Carnahan, J. W.; Mulligan K. J.; Caruso, J. A. *Anal. Chim. Acta,* **1981**, *130*, 227.
(8)   Mohamad, A. H.; Caruso, J. A.; in *Detectors for Gas Chromatography*; Giddings, J. C.; Grushka E.; Brown, P. R., Eds.; Advances in Chromatography Vol. 26, Marcel Dekker Inc., New York, New York, 1987, pp. 191- 227.
(9)   Evans, J. C.; Olsen, K. B; Sklarew, D. S. *Anal. Chim. Acta*, **1987**, *194*, 247.
(10)   Bruce, M. L.; Caruso, J. A. *Appl. Spectrosc.*, **1985**, *39*, 942.
(11)   Sklarew, D. S.; Olsen K. B; Evans, J. C. *Chromatographia*, **1989**, *27*, 44.
(12)   Moussounda, P. S.; Rausan, P.; Mermet, J. M.; *Spectrochim. Acta*, **1985**, *40B*, 641.
(13)   Freeman, J. E.; Hieftje, G. M. *Spectrochim. Acta*, **1985**, *40B*, 653.
(14)   Pivonka, D. E.; Fateley, W. G.; Fry, R. C. *Appl. Spectrosc.*, **1986**, *40*, 291.
(15)   Haas, D. L.; Caruso, J. A. *Anal. Chem.*, **1985**, *57*, 846.
(16)   Hagen, D. F.; Marhevka, J. S.; Haddad, L. C. *Spectrochim. Acta*, **1985**, *40B*, 335.
(17)   Slatkavitz, K. J.; Hoey, L. D.; Uden, P. C.; Barnes, R. M. *Anal. Chem.*, **1987**, *57*, 243.
(18)   Perpall, H. J., Uden, P. C., Deming, R. L. *Spectrochim. Acta*, **1987**, *42B*, 243.
(19)   Dagnall, R. M.; Pratt, S. J.; West, T. S. *Talanta*, **1970**, *17*, 1009.
(20)   Chiba, K.; Haraguchi, H. *Anal. Chem.*, **1983**, *55*, 1504.
(21)   Dingjan, H. A.; deJong, H. J. *Spectrochim. Acta*, **1983**, *38B*, 777.
(22)   Chongrong Jia; Yieru Huang; Qingyu Ou *Anal. Instrum. (Chinese),* **1989**, *4*, 34.

(23) Quimby, B. D.; Delaney, M. F.; Uden, P. C.; Barnes, R. M. *Anal. Chem.*, **1979**, *51*, 875.

(24) Zerezghi, M.; Mulligan, K. J.; Caruso, J. A. *J. Chromatogr. Sci.*, **1984**, 22, 348.

(25) Beenakker, C. I. M. *Spectrochim. Acta*, **1977**, *328*, 173.

(26) van Dalen, J. P. J.; de Lezenne Coulander, P. A.; de Galan, L. *Anal. Chim. Acta*, **1977**, *94*, 1.

(27) Tanabe, K.; Haraguchi, H.; Fuwa, K. *Spectrochim. Acta*, **1981**, *36B*, 633.

(28) McLean, W. R.; Stanton, D. L.; Penketh, G. E. *Analyst*, **1973**, *98*, 432.

(29) Dingjan, H. A.; deJong, H. J. *Spectrochim. Acta*, **1981**, *36B*, 325.

(30) Slatkavitz, K. J.; Uden, P. C.; Hoey, L. D.; Barnes, R. M. *J. Chromatogr.*, **1984**, *302*, 277.

(31) Kewei Zeng; Qingyu Ou; Guochuen Wang; Weile Yu *Spectrochim. Acta*, **1985**, *40B*, 349.

(32) Brenner, K. S. *J. Chromatogr.*, **1978**, *167*, 365.

(33) Weile Yu *J. Anal. Atomic Spectrom.*, **1988**, *3*, 893.

(34) Bonnekessel, J.; Klier, M. *Anal. Chim. Acta*, **1978**, *103*, 29.

(35) Gough, T. A.; Pringuer, M. A.; Sugden, K.; Webb, K. S.; Simpson, C. F. *Anal. Chem.*, **1976**, *48*, 583.

(36) Hooker, D. B.; DeZwaan, J. *Anal. Chem.*, **1989**, *61*, 2207.

(37) Weile Yu; Qingyu Ou; Kewei Zeng; Guochuen Wang In *Proc. 4th International Symposium on Capillary Chromatography (Hindelang/allgau), Germany, May 3-7, 1981)* Kaiser, R. E. Ed.; Huthig, Heidelberg, Germany, 1981, p.445.

(38) Qingyu Ou; Guochuen Wang; Kewei Zeng; Weile Yu *Spectrochim. Acta*, **1983**, *38B*, 419.

(39) Yieru Huang; Qingyu Ou; Weile Yu *J. Anal. Atomic Spectrom.*, **1990**, *5*, 115.

(40) Freeman, J. E.; Hieftje, G. M. *Spectrochim. Acta*, **1985,** *40B*, 475.

(41) Weile Yu In *Proceedings of Sino-West German Symposium on Chromatography, November 5- 9, 1981, Dalian, China.* Lu, P. U.; Bayer, E. Eds., Science Press, Beijing, China, 1983, pp. 38- 57.

(42) Estes, S. A.; Uden, P. C.; Barnes, R. M. *J. Chromatogr.*, **1982**, *239*, 181.

RECEIVED August 6, 1991

Chapter 4

Characterization of Interferences Affecting Selectivity in Gas Chromatography—Atomic Emission Spectrometry

James J. Sullivan and Bruce D. Quimby

Hewlett-Packard Company, Route 41 and Starr Road, Avondale, PA 19311

Atomic emission spectrometry used for gas chromatographic detection (GC-AES) has unique capabilites for solving chemical and spectral interferences. Chemical interferences are minimized by water cooling of the discharge tube, by any of 3 reagent gases added to the plasma, and by controlling the makeup flow. Chemical interferences with S, O, N, P, Si, H, D, and F are discussed. Spectral interferences with chromatographic detection are corrected using detecting algorithms called recipes. These perform real-time multipoint background correction with matched filters for the signal and background spectra. Examples of recipes for S, Cl, Br, and ^{13}C are shown. Snapshots (spectra collected during chromatograms) are used to confirm elemental determinations in individual compounds. Spectral interferences with snapshots are removed by novel software techniques. Snapshots are shown for low-levels of S, Cl, P, H, ^{13}C, and metals in various samples including crude oil and pesticides.

Atomic emission spectrometry provides elemental selectivity for gas chromatography (GC-AES), but chemical and spectral interferences can reduce selectivity. A new atomic emission detector (AED), which has reduced these interferences, has been described in detail (1, 2). Applications of the AED have been developed for C, H, N, and O (3), pesticides and petroleum samples (4), empirical formulas (5) and screening applications (6). The technique has evolved considerably from that described in earlier work (7-18).

In GC-AES, there is sometimes a chemical interaction between the element of interest and the wall of the discharge tube. If insufficient reagent gas is used, carbon can be deposited as graphite; other interactions can liberate a species adsorbed on the wall of the tube, and a fluorine-containing compound will remove silicon from the tube wall.

Spectral selectivity is often quite high, but a spectral overlap can cause a response from an interfering chromatographic peak. For instance,

0097–6156/92/0479–0062$08.00/0

hydrocarbons lead to molecular emissions which overlay the atomic lines of sulfur. This causes an undesired response on the sulfur chromatogram, and a reduction in sulfur selectivity. In GC-AES, "snapshots" (real time spectra) are collected during a chromatogram and used to confirm the presence of an element in a peak. The snapshots show spectral interferences due to molecular bands, varying with the chromatographic profile, which can obscure the atomic lines of interest.

This article will discuss three strategies for minimizing these interferences. Chemical interferences in the AED are handled with reagent gases (7-9, 15) which are added to the plasma to prevent carbon deposition or, in some cases, to avoid refractory oxide formation. A helium makeup gas provides rapid flow through the plasma to minimize wall reactions. Water cooling of the discharge tube lowers the wall temperature drastically, thereby lowering the reactivity of the wall.

In most cases, when a spectral interference reduces chromatographic selectivity, the interfering species is known. The "recipe", or algorithm for detecting the chromatographic output of an element, can correct for this. In snapshots, spectral interferences need to be removed, so that atomic lines can be identified and displayed graphically. Two techniques to do this are background subtraction, which removes a static spectral interference, and background stripping, which removes a time-varying interference.

Experimental Section

The Hewlett-Packard (Avondale, PA) 5921A atomic emission detector was used with the HP 5890A gas chromatograph. The system, controlled by a HP5895A ChemStation, is fully automated to analyse 100 samples for many elements in each chromatographic peak. An HP 7673A automatic injector was used. The ChemStation includes a 332 series computer, a 40 megabyte disc drive, and a PaintJet printer. The GC-AED software runs on a Pascal operating system. An Eventide WPB-109 "print buffer" was used to speed data output.

Helium was purified by passage through a heated catalytic purifier (SAES Getters, Colorado Springs, CO). The gas chromatograph was coupled to the microwave plasma by extending the column through a heated transfer line to the interior of the discharge tube, the end of the column being positioned 1 cm from the plasma.

Plasma. The plasma source is a descendent of the reentrant cavities developed for use with klystrons (19). Like the Beenakker cavity (20), it is circularly symmetric, with the most intense electric field along the axis of the centrally positioned discharge tube. Fifty W of microwave power support a plasma in a helium atmosphere. Figure 1 shows a cross section of the microwave cavity. The 1 mm diameter discharge tube has a very thin wall (0.12 mm thick) which is cooled by a flow of water, which greatly reduces reactions with the wall.

Others have attempted to solve the wall reaction problem with a tangential flow torch (TFT) (21-28). In this approach, a large diameter discharge tube is used, and a very high flow of sheath gas forces the analyte

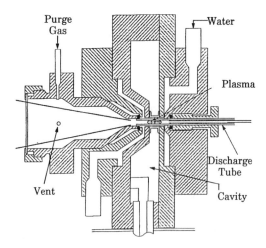

Figure 1. Cutaway view of the microwave plasma source.

flow to remain in the center of the discharge, away from the wall. The TFT was not pursued here, because the high flow of helium is much more expensive, and sensitivity suffers when the effluent is diluted by the high flow. It was considered that better overall performance would be achieved by controlling the chemistry of the discharge tube wall.

For the plasma used here, three different reagent gases controlled chemical interferences. A 45 mL/min makeup flow of helium minimized the flow time through the plasma. With certain elements, the makeup flow was increased to 150-200 mL/min, to reduce the interactions with the wall. Auxiliary flows were introduced into the discharge tube along the outside of the chromatographic column. The effluent from the plasma flowed into a chamber enclosed by a window; this configuration prevented ambient air from back-diffusing into the plasma.

Spectrometer. Figure 2 is a schematic view of the spectrometer. It is based on a concave holographic grating, is purged with nitrogen, and operates from 160 to 800 nm. The resolution of the spectrometer is 0.1 nm at 400 nm. The optical throughput of the spectrometer is determined by several components. An elliptical condensing mirror, at F/1.8, is focused along the axis of the discharge tube. The F/3.6 spectrometer has an entrance slit 50 um wide. The height of the entrance slit is vignetted by the 1 mm inside diameter of the discharge tube. The efficiency of the grating has a 30 percent maximum in the near UV.

Photodiode Array. The detector is a silicon photodiode array (PDA), 12.5 mm long, with 211 pixels, or detecting elements (*29*), and has a quantum efficiency between 40 and 80 percent. The PDA can be positioned anywhere along the 350 mm length of the focal plane of the spectrometer.

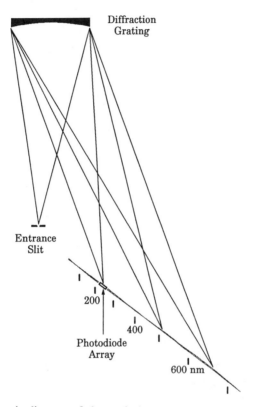

Figure 2. Schematic diagram of the optical spectrometer. (Reproduced from ref. 1. Copyright 1990 American Chemical Society.)

The length of the PDA determines which elements can be detected simultaneously. It also sets the span of the snapshots stored during a chromatogram. For instance, at 180 nm, the PDA spans 44 nm, allowing simultaneous detection and snapshot coverage of nitrogen, sulfur, and carbon, as well as some metallic elements. Near 490 nm, the PDA spans 27 nm, so that simultaneous chromatograms of H, C, Cl and Br can be detected.

Table I shows typical results. Detection limits are near one pg/sec for many elements, and selectivity (response ratio for equal weights of carbon and heteroatom) ranges from 3 to 6 orders of magnitude.

Table I. **Typical GC-AES Results**

Element	Wave-Length	MDL (pg/sec)	Selectivity vs. Carbon	Lin. Dynamic Range	Reference
C	193.1	0.2		10 k	2
H	486.1	1		9 k	2
D	656.1	2	300 (vs H)	10 k	2
O	777.2	50	30 k	3 k	2
N	174.2	15	5 k	10 k	2
S	180.7	1	20 k	10 k	2
P	177.5	3	8 k	1 k	2
Cl	479.5	25	10 k	10 k	2
Br	478.6	30	6 k	1 k	2
F	685.6	60	50k	2k	2
I	184.4	20 [b]	15 k [b]		
^{13}C	342.8	7	2 k	1 k	6
Si	251.6	3	15 k	1 k	2
Hg	253.7	0.1	300 k		4
Sn	303.4	1	37 k		6
As	189.0	3	39 k	1 k	a
Se	196.1	4	50 k	1 k	a
Sb	217.6	5	19 k	1 k	a
Ni	301.2	1	200 k	1 k	a
V	292.4	4	36 k	1 k	a
Cu	324.7	1	100 k	1 k	a
Co	345.3	3	60 k	1 k	a
Fe	302.1	0.05	1000 k	1 k	a
Zn	213.9	1 [b]	30 k [b]		
Pb	405.7	1 [b]	30 k [b]		

[a]B. D. Quimby and J. J. Sullivan, Pittsburgh Conf., New York City,
   March 5-9, 1990, Paper # 465.
[b]estimated

Chemical Interferences

Chemical interferences can be handled by excluding the interferent, depleting the interferent by a competitive reaction, or buffering the interferent. Reagent gases added to the plasma offer powerful ways of accomplishing these goals. Like other parts of a recipe for an element, these gases are turned on automatically when the element is run.

Early Work with Sulfur. The development of this AES system began with work on carbon, nitrogen and sulfur, which are detected most sensitively in the vacuum ultraviolet (VUV) (17, 18). With the PDA-based spectrometer, these three elements can be measured simultaneously. When the development began, a Beenakker cavity was used with a conventional discharge tube. An oxygen reagent was used to prevent carbon deposits. Sulfur (and nitrogen) tailed noticably more than carbon for the same peak. While little mention is made of tailing in the literature, careful inspection of the chromatograms in early work shows excessive tailing of nitrogen and sulfur as compared to carbon (*9*).

Sulfur tailing increases with increased temperature of the discharge tube and the relative degree of etching visible on it. With the traditional discharge tubes used in GC-AES, that is, thick-walled uncooled tubes, tailing is noticable even with a new discharge tube. Etching proceeds rapidly, so that sulfur detection is seriously degraded after 20 to 30 hours of operation.

As an intermediate step in the development of the water-cooled discharge tube, traditional discharge tubes were cooled with several liters per minute of air flow. To facilitate cooling, the tube wall was made thinner (1 mm wall thickness). With cooler walls, the sulfur tailing and etching problems decreased, and silicon emission (derived from the wall of the tube) also decreased. However, the improvements were not enough to solve the problem.

At this point, the water-cooled discharge tube, with 0.12 mm thick walls, was developed. The microwave tuning was changed significantly by the presence of water, so that a new cavity design was necessary. With the water-cooled, thin-walled discharge tube, the etching was greatly reduced, the silicon emission was not detectable, and the sulfur peak shape was significantly improved. The lifetime of a discharge tube, as measured by etching and sulfur tailing, was increased to several months of continuous operation. As Table I shows, performance with sulfur is substantially improved over previous results (*7-9, 13-15*).

An unexpected benefit of water-cooled tubes was decreased carbon background. Presumably, in the uncooled silica tubes, materials such as carbonates were being liberated at high temperatures.

Oxygen Detection. Until recently, attempts to analyse oxygen at the 777.2 nm line met with much less success than with other elements (*7-11,25, 30-31*). The detection limits were usually in the ng/sec range, and the highest selectivities were only a few hundred. However, better results have recently been reported (*32*).

In early work in this laboratory, the results for oxygen were unsatisfactory, since hydrocarbons gave an oxygen response by liberating small amounts of oxygen from the discharge tube walls. Figure 3a shows a chromatogram of a mixture of roughly equal levels of 3 oxygenates and 6 non-oxygenated compounds. The undesired response of the non-oxygenated compounds was due to atomic oxygen emission, as confirmed by snapshots. This indicates that a wall reaction was involved.

It was found that this wall interaction could be buffered successfully by using a reagent gas containing a few percent methane or propane in nitrogen and/or hydrogen. Nitrogen or hydrogen is needed to prevent carbon condensation. A good mixture is 0.03 mL/min hydrogen and 0.25 mL/min nitrogen/10% methane. The levels of hydrocarbon and the other components of the reagent gas are more critical that with other applications of the AED. Excess hydrocarbon depresses the response of oxygen-containing compounds.

Insufficient nitrogen or hydrogen, relative to the added hydrocarbon, leads to carbon deposition on the walls. The noted reagent gas mixture allows practical and routine detection of oxygen, and prevents carbon deposition within the discharge tube. However, the reducing nature of this gas allows carbon particles to form in the plasma exhaust, which gradually reduces the transmission of the window of the spectrometer.

This mixed reagent gas was used for the oxygen chromatogram as shown in Figure 3b. While the non-specific interferences are gone, there is tailing of the oxygen-containing peaks. Further optimizing of the reagent, and a new discharge tube reduced the tailing by a factor of 2.5.

Oxygen has been found to vary in response factor among different compounds. In compounds containing both silicon and oxygen, the oxygen response factor was found to be depressed by 25 percent (3).

Nitrogen. For nitrogen detection, the use of oxygen as a reagent gas presents two problems (2). There is much more tailing for the nitrogen response than for the carbon response, and more seriously, there is a non-specific response to all hydrocarbons, similar to that for oxygen in Figure 3a.

Snapshots taken during the tail of the nitrogen-containing peak, as well as during the other peaks, showed that the undesired responses were due to emission from atomic nitrogen, and not from molecular spectra or continuum shifts. This implies that there is an interaction between nitrogen and the wall of the discharge tube and that the slow release of nitrogen from the wall causes tailing. Peaks not containing nitrogen displace the nitrogen bound to the wall, causing the nonspecific response. With a hydrogen reagent gas, the response was unusable. The nonspecific response becomes partially negative, and the tailing time constant increases to more than a minute.

It was a pleasant surprise, therefore, to find that a combination of oxygen (0.15 mL/min) with a lower amount of hydrogen reagent (0.03 mL/min) showed improved performance (2). Nitrogen tailing was not detectable, and the selectivity was improved substantially. The detection limit, selectivity, and dynamic range for nitrogen are substantially improved over previous results (Table I).

Phosphorus. Like carbon, sulfur and nitrogen, phosphorus is also detected in the VUV region, but it cannot be run simultanously with these elements, because a hydrogen reagent gas is required. With oxygen, or oxygen and hydrogen reagent gas, even a large level of phosphorus (1 ng) gave no detectable response. It appears that phosphorus reacts with oxygen to form involatile compounds that adhere to the wall (15). In fact, changing to a hydrogen reagent gas after a phosphorus-containing peak has eluted will cause a noticable release of phosphorus. Even with hydrogen reagent gas, phosphorus has severe tailing with the traditional uncooled discharge tubes, which was reduced but not eliminated with the cooled discharge tube. However, when the makeup gas was increased from 30 to 150 mL/min, the residual tailing was small. Figure 4a shows the carbon channel of a test mixture containing triethyl phosphate; because the compound is polar, some tailing on the chromatographic column is evident. Figure 4b is the phosphorus chromatogram

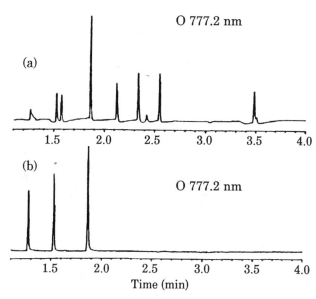

Figure 3. Oxygen chromatograms of 3 oxygenated and 6 non-oxygenated compounds, with different reagent gases: (a) hydrogen; (b) hydrogen-methane-nitrogen.

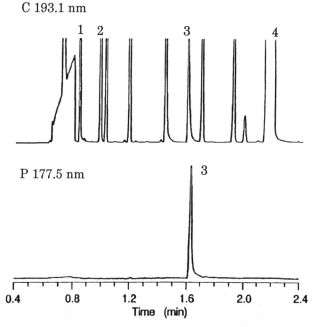

Figure 4. Carbon and phosphorus chromatograms of test mixture. Marked peaks are: (1) 3.6 ng of tetravinylsilane; (2) 5.8 ng fluoroanisole; (3) 2.4 ng of triethyl phosphate, (4) 390 ng n-decane.

with hydrogen reagent and 150 mL/min makeup gas. The phosphorus response tails only slightly more than the carbon channel.

The detection limit, selectivity and dynamic range for phosphorus are listed in Table I. The tailing mechanism also degrades the linearity of phosphorus response, which varies by 30% over its range of 3 decades. However, calibration plots are very repeatable.

Silicon. Until the development of the water-cooled, thin-walled discharge tube, silicon detection was not very practical because of the high background from the tube wall. Silicon results have been reported, but only with extremely high makeup flows (450-620 ml/min) (*13,15*). These high flows lowered the silicon background by dilution and/or by cooling the interior walls of the discharge tube. With the cooled tube, there is no detectable silicon background with the oxygen reagent gas, even at makeup flow rates as low as 20 mL/min. There is still some silicon emission, however, with hydrogen reagent gas. While hydrogen alone is unsatisfactory, the oxygen and hydrogen reagent mixture gives improved selectivity and sensitivity.

With the cooled discharge tube, a high makeup flow is no longer needed. With a makeup flow rate of 40 ml/min, the detection limits are comparable with previous results (7,9), but selectivity is dramatically increased (Table I). Silicon responds over 4 decades, but its response factor varies by +/- 20% over this range. With a fivefold higher makeup flow, both sensitivity and selectivity improve by a factor of 5 (*2*). Figure 5 shows a silicon chromatogram of a test mixture. The silicon peak shape is excellent, with no detectable tailing. The small peak at 0.94 min is a fluorine compound, fluorine attacking the discharge tube and liberating silicon (*7,8*).

Deuterium and Hydrogen Detection. Hydrogen can be detected at 486 nm (simultaneously with C, Cl and Br) or at 656 nm together with deuterium; oxygen is used as reagent. At 486 nm, hydrogen often shows tailing, sometimes in excess of the carbon channel. The tailing of hydrogen is a symptom of a larger problem. Some workers have observed nonlinear response for hydrogen (*33*), with a response factor increasing with concentration by a factor of two (20 to 30 % per decade) over the range of response (*5*). Other workers, however, have found linear response to hydrogen for more than 3 decades (*34*). Presumably a wall reaction is involved where nonlinearity has been observed, but the details are not yet understood. Also consistent with the idea of a wall effect, is a hydrogen interference seen with highly polychlorinated compounds (*5*). Work is continuing on ways to correct this interference.

For hydrogen and deuterium detection near 656 nm, oxygen reagent gas (0.15 mL/min) and an order sorting filter (cutoff at 460 nm) are used. A high makeup flow (200 mL/min) is needed to minimize deuterium tailing. Curiously, deuterium can tail more than hydrogen, but a high makeup flow reduces this to a neglegible level. A figure in *Reference 9* also shows deuterium with more tailing than with carbon.

The detection limit for deuterium is good, and the selectivity appears to be the highest yet reported (*7-9,15*).

Fluorine. Fluorine tends to tail and lose response due to reactions with the discharge tube wall, with either oxygen or hydrogen reagent gas alone. However, with a mixed reagent of hydrogen and oxygen, the sensitivity, peak shape, and selectivity are improved, though a small degree of tailing is evident. Fluorine also exhibits some curvature in its calibration curve, with the maximum deviation from linearity being approximately 20%.

The purpose of reducing or eliminating the interference problems is to enhance chromatographic selectivity. The selectivity of the AED is compared to that of an electron capture detector (ECD) in Figure 6. The sample is an extract of onions spiked with eight pesticides, most of them at 0.2 ppm. Not all of these pesticides are sufficiently sensitive on the ECD and there are many natural components in onions that also respond. By contrast, the AED can detect the pesticides on several elemental channels with no noticable inteferences, and with all of the target compounds detected.

Quantitative Measurements of Chemical Interferences

Most chemical interferences found so far with this AED involve the wall of the discharge tube, the most evident symptom of this problem being tailing of the chromatographic peak. Therefore, the degree of tailing is a good measure of the interaction (2). When a peak tails, the width at the base of the peak increases faster than the peak width measured at half maximum. Following chromatographic convention, the tenthwidth (the peak width at one tenth of peak maximum), is used to measure tailing.

A tailing process can be modeled as a gaussian chromatographic peak, convolved with an exponential broadening process, the latter being characterized by a time constant, t. Since a tailing process is more evident on a narrow peak than a wide one, increases in peak width due to tailing depend both on t and the unbroadened peak width.

Figure 7 shows the change in tenthwidth as a function of time constant for a gaussian peak with a standard deviation of one second. The halfwidth of the unbroadened peak is 2.36 seconds (FWHM), and the unbroadened tenthwidth, W_0, is 4.30 seconds. Given measurements of tenthwidth, W, for a broadened peak, the equivalent time constant can be found from the graph.

For low time constants, the time constant, t, can be fitted to the following curve based on the squares of the broadened and unbroadened tenthwidths.

$$t = 0.249 * \sqrt{W^2 - W_0^2}$$

When a compound is chromatographed, the carbon-specific peak usually shows neglegible broadening, so it can be used as the unbroadened reference. The tenthwidth is measured for the other elements in the same compound, and the equivalent time constants are calculated. For typical chromatographic peaks, a time constant equal to the halfwidth can be considered as approximately the threshold between negligible tailing and serious tailing. The

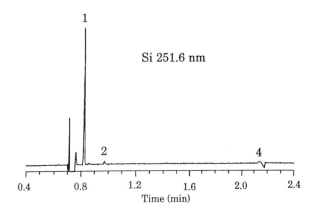

Figure 5. Silicon chromatogram of same mixture as Figure 4.

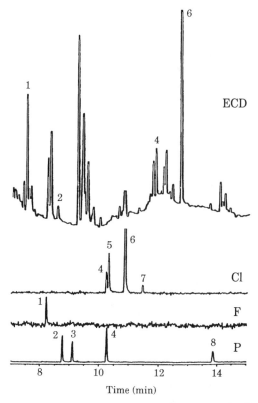

Figure 6. Pesticides in extract from onions on ECD and AES (Cl, F, and P). All compounds are at 0.2 ppm, unless noted. Peaks are: (1) 1 ppm ethalfluralin, (2) dimethoate, (3) diazinon, (4) chlorpyrifos and parathion (unresolved), (5) chlorthal-dimethyl, (6) 5 ppm folpet, (7) dieldrin, and (8) azinphos-methyl. (Reproduced with permission from reference 6. Copyright 1991 Elsevier Science Pub.)

halfwidths of peaks depend on column efficiency, injection volume, column length and retention time, but a typical peak on a capillary column has a halfwidth close to 1 second.

Tailing time constants vary with the element, type of reagent gas, and the relative concentrations of compound and reagent gas. To investigate the effect of the chemical nature of the discharge tube, measurements were made both on the standard silica tube and on an experimental sapphire tube.

Table II shows the measured time constants for several elements, reagent gases, and tube materials. The listed reagent gases are those normally used with a silica discharge tube, except for the last entry, phosphorus with oxygen reagent gas. Of the elements listed, only oxygen, fluorine and phosphorus have measurable tailing on silica.

Table II. **Tailing Time Constants**

Element	Reagents	Makeup Flow (mL/min)	seconds Silica Tube	Sapphire Tube
Chlorine	O_2	30	nd[a]	nd
Bromine	O_2	30	nd	1.0
Hydrogen	O_2	30	nd	nd
Deuterium	O_2	150	nd	0.8
Nitrogen	O_2 H_2	40	nd	nd
Silicon	O_2 H_2	40	nd	nd
Sulfur	O_2 H_2	40	nd	0.5
Oxygen	H_2 N_2 CH_4	40	0.6	0.9
Fluorine	H_2	150	0.4	15.6
Phosphorus	H_2	150	1.1	1.8
Phosphorus	O_2	150	lost	nd

[a]nd means time constant less than 0.2 seconds

Based on Table II, the silica discharge tube seems to be the better overall choice. While fluorine has less tailing on silica than on sapphire, it some cases, sapphire may be a better choice for applications containing fluorinated samples. This is because fluorine liberates both silicon and oxygen from silica, but not from sapphire. When silicon or oxygen determinations are required for compounds containing fluorine, the fluorine interference is greatly reduced with a sapphire discharge tube, at the expense of severe tailing of the fluorine response, and lower amounts of tailing of other elements.

Table III lists tailing time constants for silica and sapphire with both hydrogen and oxygen reagent gases. These results were measured with water-cooled discharge tubes which had a thicker tube wall (1 mm), so that the inside surface had a higher temperature. The tailing is generally worse than for the thinner walled tubes. For instance, chlorine with oxygen reagent has a measurable time constant of 0.4 sec. Phosphorus is much worse with both types of discharge tubes. Makeup flow is 40 ml/min.

Table III. **Tailing Time Constants with Various Reagents**

| | seconds | | | |
| | Silica Tube | | Sapphire Tube | |
Element	O_2	H_2	O_2	H_2
Carbon	nd	6	0	0.8
Nitrogen	2.9	lost	0.8	nd
Chlorine	0.4	1.4	1.3	8.4
Deuterium	0.2		0.3	8.4
Hydrogen	nd		nd	
Sulfur	2.6	7.5	23.1	0.4
Oxygen		lost		lost
Phosphorus	lost	12	lost	lost

Table III shows that other combinations of reagent gases can sometimes be used, with various increases in tailing. For instance, with the silica tube, sulfur showed a moderate tailing with oxygen, but had a 7.5 second time constant with hydrogen; chlorine had a 1.4 second with a hydrogen reagent.

It is interesting to note that, with the silica tube, phosphorus tails less with a hydrogen reagent, but with a sapphire tube, oxygen reagent gives the least tailing. With sulfur, the opposite is true: oxygen works best with silica, but hydrogen works best with sapphire. With chlorine, the oxygen reagent gas gives lower tailing than hydrogen on both types of discharge tube.

Spectral Interferences in a Chromatogram

Figure 8 illustrates the selectivity problem which exists when element-selective chomatograms are made. Figure 8a shows a shapshot taken during the elution of a sulfur-containing peak, and Figure 8b shows a snapshot during a hydrocarbon peak. The response is plotted as a stair-step. Each step shows the response from one pixel, or detecting element, of the PDA. Sulfur has three emission lines, which produce response on 6 pixels which unfortunately all have response from hydrocarbons and other peaks not containing sulfur. A sulfur-selective chromatogram could be made by monitoring these 6 signal pixels, but it would not be very selective, because there is some response to sulfur-free compounds. Background correction can compensate for the undesired "background" response of sulfur-free compounds. An interferent such as that shown in Figure 8b, has similar levels of response on the signal pixels and on nearby pixels. If a calibrated response from some nearby pixels is subtracted from the response of the signal pixels, the response to the interferent can be nulled.

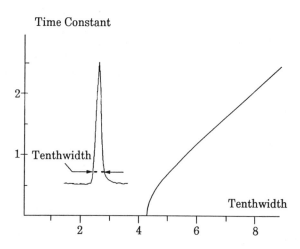

Figure 7. Exponential time constant, versus tenthwidth of broadened gaussian peak.

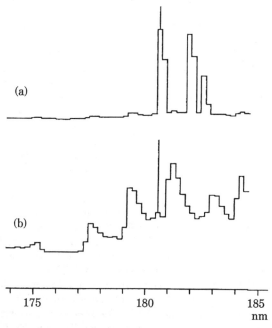

Figure 8. Snapshots collected during elution of (a) sulfur-containing compound, (b) hydrocarbon.

Recipes for Chromatographic Detection. A recipe is the software algorithm used to make an element-selective chromatogram. It is made up of two digital filters, which operate on the set of pixel responses to detect the raw atomic emission from the element and the interferent.

The structure of the two filters is shown in Figure 9. Each pixel signal is scaled by its individual gain factor (triangles). The scaled pixel signals are then summed together, forming the two filter outputs. Mathematically, each of the outputs from the two filters is the scalar product, $<A,S>$, of a filter, A, and the set of PDA outputs, S, that is, the sum of the gain terms, A_i, multiplied by the corresponding pixel signals, S_i, for each pixel i. Chromatograms are formed by recording the outputs of the raw signal filter and background filter, and both are stored in memory.

Optimal Recipes. The pixels of the signal filter are chosen at positions where the atomic emission is present. There is an optimal set of gain terms for the signal filter; when the noise level is the same on each pixel, as is usually the case with small signals, the best choice of gain terms is a matched filter (35,36). In a matched filter, the plot of the gain terms has the same shape as the spectrum of the signal. The two filters for a recipe are made by selecting portions of snapshots of desired elements and of interfering responses, as shown in Figure 8. In this way, matched filters are automatically obtained.

The two stored chromatograms for an element record the raw elemental output, $<E,S>$ and the background output, $<B,S>$, from an elemental filter, E, and a background filter, B. The complete chromatogram is a linear combination of these: $<E,S> - k<B,S>$. The factor k by which the background chromatogram is scaled before being subtracted from the raw elemental chromatogram is termed the "background amount". The value of k can be set experimentally, by selecting a chromatographic peak, M, which is free of the element of interest. During this peak's elution, the values of the raw elemental response and the background response are measured. The ratio of these responses, $<E,M>/<B,M>$, is the background amount, k.

Figure 10 shows chromatograms of a sulfur-selective recipe. The mixture contains 9 compounds, the first 8 all being similar in concentration, but the ninth being two orders of magnitude more concentrated. Only the seventh peak contains sulfur. Figure 10a is the raw elemental response; while there is some selectivity to sulfur, all of the other peaks have visible response. Figure 10b is the background chromatogram, with non-specific response. The complete chromatogram, Figure 10c, is a linear combination of the other two, with an optimal value of background amount. Only the sulfur response is evident. The response factor for carbon, as measured from the last peak, is lower than the response factor for sulfur by more than a factor of 10,000.

A severe test of the recipe structure is detection of molecular species. Figure 11 shows the 0-3 and 1-4 bands of the fourth positive system of CO. Two spectra are shown: ^{12}CO and ^{13}CO. The spectra are separated enough to be distinguished. The question is, what selectivity can be achieved under chromatographic conditions. Recipes were made for ^{12}C and ^{13}C from these spectra (6, 37). On a test mixture, selectivity of 2,500 for ^{13}C-spiked compounds over unspiked compounds was achieved. The recipe was adjusted

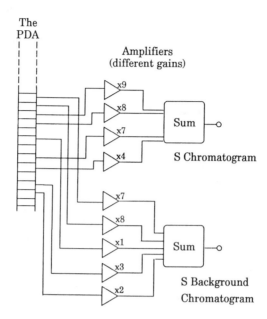

Figure 9. Structure of a sulfur recipe. PDA pixel responses, with different gains, are summed into signal and background chromatographic outputs.

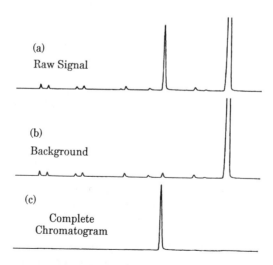

Figure 10. A complete chromatogram, (c), is made by subtracting an amount of the background chromatogram, (b), from the raw signal, (a).

to reject compounds with the natural abundance (1.1%) of ^{13}C, since otherwise, the selectivity for spiked compounds could not exceed 100.

Doubly Orthogonal Recipes. The recipe structure just described can be made orthogonal to any interference, by proper selection of the background amount. It would seem therefore, that, once the background amount was properly set, the response should be essentially zero for all compounds which exhibit the same interfering spectrum. In fact, deviations from perfect selectivity are seen even for a homologous series of interfering compounds.

Figure 12 shows chlorine- and bromine-selective chromatograms for a test mixture. The vertical scale was adjusted to make the chlorine and bromine peaks ten times full scale. There are four aliphatic hydrocarbons present in this mixture. Octane (C_8) and dodecane (C_{12}) are present at high levels, while tridecane (C_{13}) and tetradecane (C_{14}) are at much lower levels. In each chromatogram, the high level alkanes have positive responses, indicating that the background amount is too low, but the low level alkanes have a negative response, indicating that the background amount is too high. It appears, therefore, that the interfering spectra are not the same for all compounds.

Figure 13 shows the three dimensional snapshots collected during the elution of octane (C_8) and tridecane (C_{13}). The shapes of the spectra are different, with differences in relative band intensities, and probably a larger continuum component for octane. This is an example of the interfering spectrum changing shape as a function of concentration. The other two hydrocabons are present at an intermediate concentration (C_{12}) and at a lower concentration (C_{14}). Their spectra also show shape changes with concentration.

A type of recipe which is doubly orthogonal is needed; that is, a type of recipe which has null response to two different background spectra. Such a recipe whould then have improved selectivity over the whole range of concentrations of interferences. The structure of a doubly orthogonal recipe is a generalization of the simpler recipes discussed so far. A doubly orthogonal recipe has a signal filter E, but the background filter is split into two parts, B and C. Since the two parts B and C generally have different sensitivities for two interfering snapshots, there is a linear combination of B and C which can reject both interferences. Doubly orthogonal recipes were made for chlorine and bromine, using shapshots from Figure 13 to make the background filters. Figure 14 shows the output of these recipes with the same sample and conditions as Figure 12. Selectivity is improved, and the large alkane peaks have been suppressed by more than a factor of 10.

Comparison of Recipes. In an earlier paper (*1*), the various types of recipes were compared for chlorine detection. The simplest possible recipe is the response of a single pixel without background correction. This gave a weight-based selectivity for chlorine versus carbon of 100. Background correction can be done with the average response of two pixels on either side of the signal pixel. This only improved the selectivity to 170, since not enough background correction was performed. When the background amount for the three-pixel recipe was changed from 0.5 to an experimentally optimized value, the selectivity increased to 2,700, at the expense of some degradation in sensitivity.

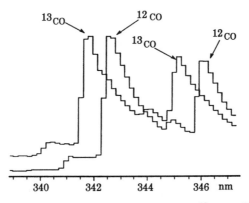

Figure 11. Snapshots of molecular spectra of CO for ^{12}C and ^{13}C. Bands are in second order, 170 - 173 nm.

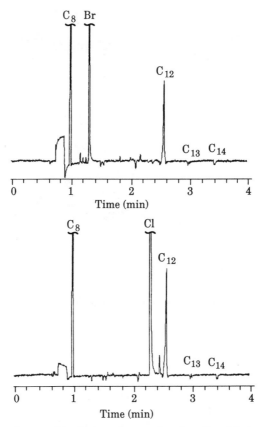

Figure 12. Test mixture with singly orthogonal recipes for chlorine and bromine. Halogenated peaks are 10 x offscale. Labeled peaks are: (C_8) 158 ng octane, (Br) 2.6 ng bromohexane, (Cl) 3.3 ng trichlorohexane, (C_{12}) 170 ng dodecane, (C_{13}) 17 ng tridecane, (C_{14}) 5.1 ng tetradecane.

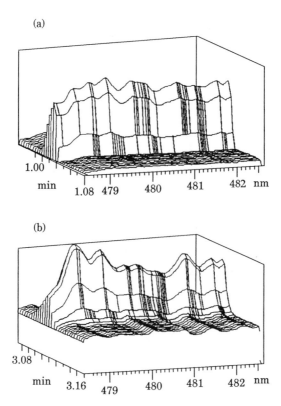

Figure 13. Snapshots taken during alkane peaks of different amounts: (a) 790 ng octane, (b) 85 ng tridecane.

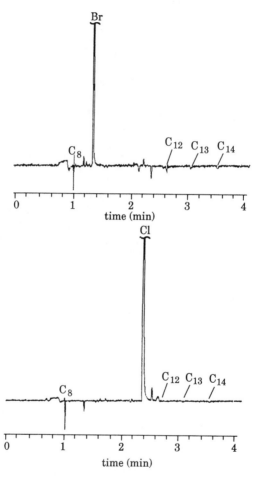

Figure 14. Test mixture with doubly orthogonal recipes for chlorine and bromine. Same sample and conditions as Figure 12.

In contrast to this, the singly and doubly orthogonal recipes based on matched filters had selectivities of 7,300 and 10,100, respectively. Both recipes had better sensitivity than the three simple recipes. The doubly orthogonal recipe preserves its high selectivity over a larger range of concentration of interferences.

Confirming Elemental Determinations. Even with selectivities exceeding 10,000, there can be artifacts in chromatography. If peaks in a run are suspected of being artifacts, the background amount, which determines the degree of chromatographic background correction, can be modified.

Figure 15 shows the sulfur chromatogram of a flavor extract of whisky. In Figure 15a, an optimal value of background amount has been used. There are several small responses just above the baseline noise, that might be interpreted as sulfur response. To check this, the background amount has been increased by about 3 percent in Figure 15b. Peaks which are not due to the element of interest often have large but cancelling components on the signal and background channels. Therefore their response changes rapidly with a small shift in the background amount, while true peaks are hardly affected.

In Figure 15b, several small reponses turn into large negative peaks, when the background amount is shifted by a small amount. This indicates that these peaks are unlikely to result from sulfur emission. The peaks marked by an asterisk change only slightly with the shift in background amount. They were confirmed as containing sulfur using the snapshots collected during the run. Each marked peak was found to exhibit the three characteristic sulfur lines near 180 nm, the levels of sulfur being between 3 and 30 pg in each peak.

Variation of the background amount is a useful tool for surveying a large number of peaks in an element-specific chromatogram. The snapshots during each peak provide positive confirmation of an elemental determination. More examples of confirmation with snapshots are given in the next section.

Spectral Interferences in Snapshots

The atomic emission detector can confirm its own elemental determinations by displaying the snapshots before, during and after a chromatographic peak. An element is confirmed if its characteristic atomic lines are present at the proper wavelengths, and if the lines have the proper relative intensities. A convenient way to evaluate these spectra is with a three-dimensional (3D) display of snapshots.

Background Subtraction. Figure 16 shows a 3D display of snapshots during a small peak containing 80 pg of sulfur. The characteristic sulfur triplet at 180-182 nm is clearly evident, and confirms the presense of sulfur in this compound. The snapshots also show a very broad band due to molecular oxygen, with a maximum at 186 nm, and many sawtooth-shaped bands due to CO. If the concentration of sulfur were much lower, it would be harder to distinguish the sulfur triplet from these background spectra. When an interfering spectrum is constant with time, it can be removed from the display by background subtraction. That is, one of the snapshots just before or after the compound

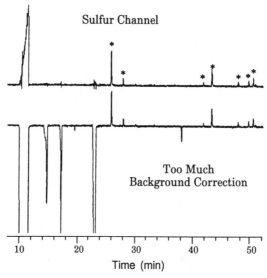

Figure 15. Sulfur chromatograms of extract of whisky flavors, with two values of the background amount. Sulfur peaks marked with "*".

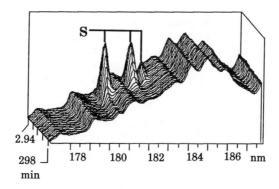

Figure 16. 3D snapshot during peak with 80 pg of sulfur. No background correction.

elutes is subtracted from all the snapshots displayed. In Figure 17, the constant background spectrum shown in Figure 16 has been removed by background subtraction.

Background Stripping. Figure 18a shows the snapshots around a characteristic chlorine triplet. If the concentration of the compound were lower relative to the interfering peaks, the chlorine peaks would not be evident, due to the background. The background spectrum has a constant shape, but the intensity varies. It is visible at 0.72 minutes, drops to a minimum in the middle of the display, and rises to a maximum near 0.80 minutes. Background subtraction would be of limited use here, since the background spectrum varies with time. It would be useful to remove the background interferences even though the background is varying. Intuitively, it seems that this should be possible, since the shape of the background is known, and is very different from the atomic lines.

Attempts were made to calculate the amount of background to subtract from each picture, by minimizing the root mean square of the remaining intensity. The results were unsatisfactory; there was always excessive background subtraction, so that the molecular bandheads became negative excursions. Since atomic emission data cannot be negative, the software was made to limit the spectral subtraction to prevent the snapshot from going negative. This non-negativity constraint gave better results, but there were often still negative-going molecular bandheads, occuring when the snapshot had a continuum component as well as a molecular component.

It was recognised that the continuum is also a spectral interference, and that its magnitude varies with sample concentration. To cope with this interference, the constant component was also removed from each snapshot. By removing the continuum first, the removal of molecular bands does not give negative-going excursions. The algorithm worked well on simulations, but real snapshots often showed incomplete background removal, due to noise. If the snapshot and the background spectrum both had low signal intensity, a negative noise excursion in the snapshot would reduce the amount of background correction. This was solved by relaxing the non-negativity constraint and letting the corrected snapshot go a few percent negative. This prevented the inaccuracies in background correction caused by a few percent noise in the snapshot.

The algorithm to do this is called "background stripping". The first stage of the algorithm is to remove the DC components from both the set of snapshots and the spectrum of the pure interference. Next, it subtracts the largest possible fraction of the interference from each snapshot that does not produce a negative response more than 3 percent of the original spectrum.

Background Stripping Examples. For three-dimensional displays, different amounts of interfering continua and spectra are removed from each snapshot. In Figure 18b, the background spectra visible in Figure 18a have been almost completely removed by background stripping. Background stripping is also useful in two dimensional snapshots. The algorithm is often more accurate than the eye in estimating the amount of background subtraction needed.

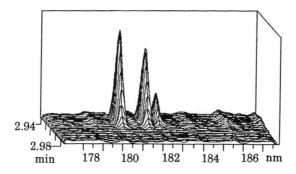

Figure 17. Same display as Fig. 16, with background subtraction.

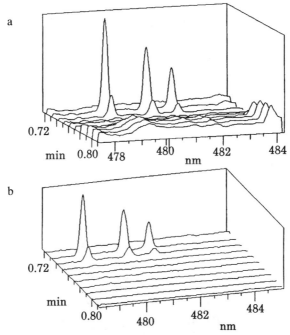

Figure 18. 3D snapshot during chlorine-containing peak. (a) no background correction, (b) background stripped.

Figure 19a shows a snapshot at the apex of a very small peak in the sulfur chromatogram, equivalent to 10 pg of sulfur. If present, the sulfur triplet would be many times smaller than the interfering hydrocarbon spectrum. Attempts to find the level of interference by trial and error were inconclusive. Figure 19b shows the same snapshot with the hydrocarbon interference removed by background stripping. All three sulfur lines are evident in the correct proportions and at the expected wavelenths. The weakest sulfur line is 50 times less intense than the interfering spectrum that has been stripped. There are a

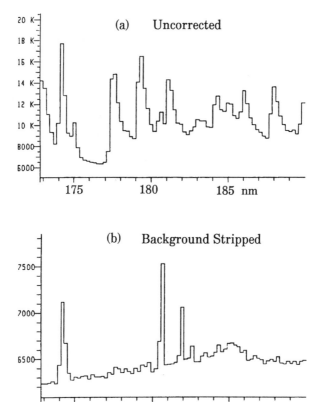

Figure 19. Snapshot during peak in gasoline containing 10 pg of sulfur. (a) no background correction, (b) background stripped.

few remaining spectral features from the molecular spectrum. The feature at 174 nm is atomic nitrogen due to a constant level of nitrogen in the plasma. The pure interfering spectrum also showed this peak, but had large amounts of the molecular spectrum. Baseline stripping scaled down the background spectrum, which caused the atomic nitrogen line to be only partially removed.

When snapshots are cleared of molecular interferences, they can be used as a routine tool for confirmation. A multi-element analysis of a broccoli extract spiked with pesticides was run and some of the spectra for a particular pesticide, chlorpyrifos, are shown in Figure 20. The spectra are characteristic for hydrogen, chlorine, sulfur and phosphorus, and confirm the presence of these elements.

Crude oils are complex mixtures which contain volatile metalloporphyrin complexes. Several of these are easily detected by AES, even in unprocessed crude oil. Figure 21 shows the snapshots obtained during a portion of a chromatogram. Since many different organometallic compounds are present for each metal, the peaks have been fused into bands, similar to the responses

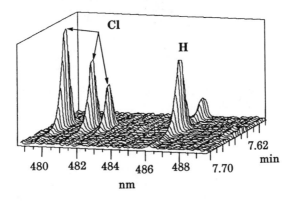

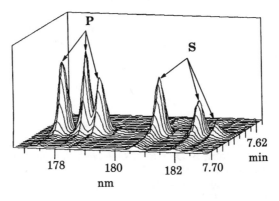

Figure 20. 3D snapshots during a chlorpyrifos peak in broccoli extract, confirming some of the elements.

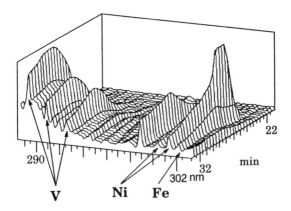

Figure 21. 3D snapshot during chromatogram of crude oil, showing porphyrins of vanadium, iron and nickel.

seen in simulated distillation chromatography. Several spectral peaks can be identified to confirm the identity of each metal. This task is made easier since atomic lines of the same element have the same chromatographic peak shape.

Conclusion. For many years, chemical and spectral interferences have severely limited the application of AES to gas chromatography. With the chemical, instrumental, and algorithmic techniques described here, these problems have been brought under control, so that the technique is now found to be practical and broadly useful in the analytical laboratory.

Literature Cited

1. Sullivan, J. J.; Quimby, B. D. *Anal. Chem.* **1990**, *62*, 1034.
2. Quimby, B. D.; Sullivan, J. J. *Anal. Chem.* **1990**, *62*, 1027.
3. Sullivan, J. J.; Quimby, B. D. *HRC CC, J. High Resolut.Chromatogr. Chromatogr. Commun.* **1989**, *12*, 282.
4. Wylie, P. L.; Quimby, B. D. *HRC CC, J. High Resolut. Chromatogr. Commun.* **1989**, *12*, 813.
5. Wylie, P. L.; Sullivan, J. J.; Quimby, B. D. *HRC CC, J. High Resolut. Chromatogr. Chromatogr. Commun.* **1990**, *13*, 499.
6. Sullivan, J. J. *Trends Anal. Chem.* **Jan. 1991**, *10*, No. 1, 24.
7. Ebdon, L.; Hill, S.; Ward, R.W. *Analyst* **1986**, *111*, 1113.
8. Uden, P. C. *Chromatogr. Forum* **1986**, *Nov-Dec.*, 17.

9. McLean, W. R.; Stanton, D. L.; Penketh, G. E. *Analyst* **1973,** *98,* 432.
10. Brenner, K. S. *J. Chromatogr.* **1978,** *167,* 365.
11. Ke-Wei, Z.; Qing-Yu, O.; Guo-Chuen, W.; Wei-Lu, Y. *Spectrochim. Acta,* **1985,** *40B,* 349.
12. Hagen, D. F.; Marhevka, J. S.; Haddad, L. C. *Spectrochim. Acta,* **1985,** *40B,* 335.
13. Quimby, B. D.; Uden, P. C.; Barnes R. M. *Anal. Chem.* **1978,** *50,* 2112.
14. Tanabe, K.; Haraguchi, H.; Fuwa, K. *Spectrochim. Acta* **1981,** *36B,* 633.
15. Estes, S. A.; Uden, P. C.; Barnes, R. M. *Anal. Chem.* **1981,** *53,* 1829.
16. van Dalen, J. P. J.; de Lezenne Coulander, P. A.; de Galen, L. *Anal. Chim. Acta* **1977,** *94,* 1.
17. Braun, W.; Peterson, N. C.; Bass, A. M.; Kurylo, M. J. *J. Chromatogr.* **1971,** *55,* 237.
18. Genna, J. L.; McAninch, W. D.; Reich, R. A. *J. Chromatogr.* **1982,** *238,* 103.
19. Hamilton, D. R.; Knipp, J. K.; Horner Kuper, J. B. *Klystrons and Microwave Triodes,* Boston Technical Publishers: Lexington, 1964; Chapter 4.
20. Beenakker C. I. M. *Spectrochim. Acta* **1977,** *32B,* 173.
21. Bollo-Kamera, A.; Codding, E. G. *Spectrochim. Acta* **1981,** *36B,* 973.
22. Haas, D. L.; Caruso, J. A. *Anal. Chem.* **1985,** *57,* 846.
23. Goode, S. R.; Chambers, B.; Buddin, N. P. *Spectrochim. Acta* **1985,** *40B,* 329.
24. Pivonka, D. E.; Fateley, W. G.; Fry, R. C. *Appl. Spectrosc.* **1986,** *40,* 291.
25. Cull, K. B.; Carnahan, J. W. *Appl. Spectrosc.* **1988,** *42,* 1061.
26. Goode, S. R.; Kimbrough, L. K. *J. Anal. At. Spectrom.* **1988,** *3,* 915.
27. Suyani, H.; Creed, J.; Caruso, J.; Satzger, R. D. *J. Anal. At. Spectrom.* **1989,** *4,* 777.
28. Heltai, G.; Broekaert, J. A. C.; Leis, F.; Tolg, G. *Spectrochim. Acta* **1990,** *45B,* 301.
29. Kamins, T. I.; Fong, G. T. *IEEE J. Solid-State Circuits* **1978,** *SC-13,* 80.
30. Zeng, K. W.; Ou, Q. Y., Wang, G. C.; Yu, W. L. *Spectrochim. Acta* **1985,** *40B,* 349.
31. Slatkavitz, K. J.; Uden, P. C.; Barnes, R. M. *J. Chromatogr.* **1986,** *355,* 117.
32. Bradley, C.; Carnahan, J. W. *Anal. Chem.* **1988,** *60,* 858.
33. Jelink, J. Th.; Venema, A. In *Proc. 11th Intl. Sympos. on Capillary Chromatography*; Sandra, P. and Redant, G., Eds., Huthig: Heidelberg, FRG, 1990; pp. 358.
34. George, M. A.; Hessler, J. P.; Carnahan, J. W. *J. Anal. At. Spectrom.* **1989,** *4,* 51.
35. Hirschfeld, T. *Appl. Spectrosc.* **1976,** *30,* 67.
36. Bialkowski, S.E. *Appl. Spectrosc.* **1988,** *42,* 807.
37. Quimby, B. D.; Dryden, P. C.; Sullivan, J. J. *Anal. Chem.* **1990,** *62,* 2509.

RECEIVED May 23, 1991

Chapter 5

Microwave-Induced Plasma–Atomic Emission Detection for Organometallic Gas and Supercritical-Fluid Chromatography
Sample Handling and Instrument Comparisons

Thomas M. Dowling[1], Jeffrey A. Seeley[1], Helmut Feuerbacher[2], and Peter C. Uden[1]

[1]Department of Chemistry, University of Massachusetts, Amherst, MA 01003
[2]Ingenieurbuero fur Analysentechnik, Tuebingen, Germany

This study considers improvements in microwave-induced plasma (MIP) atomic emission detection (AED) systems for chromatography. A new MIP torch design, the concentric dual flow torch (CDFT), which overcomes some limitations of conventional capillary quartz discharge tubes, is evaluated. Features include the ability for splitless injection of up to 1μl without extinguishing the plasma, a wide range of plasma operating conditions, a simplified background spectrum and improved spatial stability of the plasma. Sensitivities and selectivities for Sn and Se are presented. The torch is also useful for AED in capillary supercritical fluid chromatography (SFC). Applications include gas chromatographic (GC) analysis of alkyltin species in sea water with on-line hydride generation, and the detection of silicon containing compounds after separation by SFC. In addition, the features of a generally available GC-MIP system are discussed; it is shown to have excellent sensitivity and selectivity for detection of alkyltin and selenium species in complex matrices.

The use of the atmospheric pressure microwave-induced plasma (MIP) for chromatographic element specific detection has developed rapidly in the past decade or more (1-5). However, there are continuing demands for improvements in atomic emission detectors and development of new methods of analysis, which must be met before atomic emission spectroscopy (AES) can be considered a mature method of chromatographic detection. Improved hardware and a wider range of established applications will expand the utility and acceptance of AED for chromatography.

While the MIP has several features that make it an excellent gas chromatographic (GC) detector, the conventional torch design which utilizes a capillary quartz discharge tube has several limitations, the greatest of which is low sample capacity. Typical

0097–6156/92/0479–0090$06.00/0

capillary GC injection volumes (1μl split 100:1) can cause problems with plasma stability and the solvent may extinguish the plasma or cause carbon soot to deposit on the inner walls of the discharge tube. Even if the plasma can be relit, it often requires significant time to restabilize, and information present in the chromatographic region of the solvent is lost or highly distorted. Carbon deposits cause changes in spectral background, produce baseline drift and limit discharge tube lifetime.

With the conventional torch only a single plasma gas flow is used. This can give spatially unstable plasmas which may wander inside the discharge tube or remain aligned on the inner wall. Wandering causes increased flicker noise and reduced sensitivity. If the plasma remains on the wall, analyte may be swept around the plasma, also resulting in reduced sensitivity. Plasma contact with the walls makes them become much hotter, and analyte interaction with them becomes more probable, giving increased peak tailing.

Quartz discharge tubes produce a high silicon spectral background which makes silicon detection difficult and eliminates the 251.6 nm and 288.1 nm spectral regions for the monitoring of emission from other elements. As silicon erodes from the inner walls, spatial instabilities also occur. Conventional quartz discharge tubes have a narrow range of working conditions, and operating the plasma at powers above 60W or at low make-up gas flow rates drastically reduces discharge tube lifetime. This narrow range reduces the ability to tailor plasma conditions to enhance sensitivity for specific elements.

One approach to eliminating solvent problems has been to use a venting system (6-9) which directs the solvent away from the plasma while allowing the analytes of interest to reach it. Solvent venting systems have been shown to be effective, but they add complexity to the design, and information in the chromatographic region of the solvent is lost. Solvent venting does not address the problems of spatial instability of the plasma, interactions with the discharge tube walls, high silicon background and short discharge tube lifetimes.

An alternative approach is to redesign the plasma torch. Two tangential flow torch designs have been shown to overcome many of the above problems (10-11). These had good sensitivity and response linearity with improved spatial stability and solvent capacity.

A concentric dual flow torch (12) has been evaluated and the configuration shown to overcome many of the limitations of conventional MIP torches while maintaining the sensitivity and selectivity of earlier designs. Applications include alkyltin speciation in seawater and use of this torch as an AED for capillary supercritical fluid chromatography.

A GC/AED system designed and built by Hewlett-Packard, (HP5921A AED) (9,13-15) features a thin walled water cooled quartz discharge tube, a solvent venting system and a direct connection between the cavity and a nitrogen purged spectrometer. Exclusion of ambient atmosphere allows the detection of oxygen and nitrogen and the utilization of UV emission lines. Cooling of the discharge tube reduces analyte interactions with the discharge tube walls and extends its lifetime. Wavelength dispersion is achieved using a fixed grating spectrometer with a flat focal plane, and detection is performed using a movable photo-diode array. The photo-diode array detector allows custom design of matched digital filters for each emission line of interest and simultaneous multi-element monitoring within a wavelength region of about 20-30 nm. Features of this system are applied and evaluated.

Instrumentation

Concentric Dual Flow Torch (CDFT). The concentric dual flow torch (CDFT) (Ingenieurbuero fur Analysentechnik, Tuebingen, Federal Republic of Germany) was fitted to a TM_{010} microwave cavity modified so that the CDFT fittings threaded directly into the back of the cavity (Figure 1). Ceramic discharge tubes (Scientific Instruments Inc. Rigoes, NJ, 99% alumina) were used. The inner tube was 50 mm long and 2.0 mm o.d. x 1.2 mm i.d. The outer tube was 30 mm long and 4.0 mm o.d. x 2.4 mm i.d. Nichrome wire was wrapped around the inner tube and the outer tube slipped over the wire coil so that the tubes overlapped by about 1 cm. The inner tube was positioned so that the end was 1-2 mm from the plasma and the end of the outer tube was flush with the front of the cavity. Microwave energy from the generator (Raytheon Microtherm Model CMD 5, Raytheon Co., Waltham, MA.) was coupled to the cavity via a coaxial cable and tuning was achieved with a double stub stretcher and a screw in the face plate of the cavity. Operation of the torch is described below in more detail.

GC-AED with the CDFT. A 5840A gas chromatograph (Hewlett-Packard, Avondale, PA) was interfaced to the plasma via a heated transfer line. A fused silica capillary column (DB-5, 25 meter, 0.32 mm i.d., 0.2 μm film thickness, J+W Scientific, Inc., Folsom, CA) was passed through the transfer line and through the inner ceramic tube. The column was held in place (0.5 cm from the plasma) with a graphite ferrule and nut that threaded onto the end of the torch assembly. The transfer line (35 cm long) extended from the GC oven to the end of the torch where it overlapped the column connection by 0.5 cm. A prototype echelle grating monochromator (Spectraspan, Applied Research Labs, Valenica, CA) was used for wavelength dispersion. The spectrometer and detection system have been described elsewhere (16).

SFC-AED with the CDFT. The supercritical fluid chromatographic system has been described elsewhere (17). A 50 μm frit restrictor was connected to the column (20 meter, 0.2 mm i.d., 0.05 μm film thickness, J+W Scientific, Inc., Folsom, CA) with a 1/32" union (ZU.5FS.4, Valco Instrument Co Inc., Houston TX,). The restrictor was passed through a heated transfer line and connected to the torch assembly as described for the GC column. Samples were injected with an electronically actuated micro-valve (Valco) with a 0.1 μl loop. A retention gap was connected directly to the injector and to the head of the column with a 1/32" union.

No split vent was used. The image of the plasma was focused on the entrance slit of the monochromator (Heath Model EU-703, 0.35-m focal length, Czerny-Tuner scanning monochromator, McPherson Instruments, Acton, MA). This spectrometer and detection system have been described previously (8).

GC-AED with the HP-5921A. The Hewlett-Packard AED system (Model 5890 II GC with 5921A spectrometer detector) supplied with a HP-7673A auto-injector and 'Chem Station' for data aquisition and processing has been described elsewhere (9,13-15). A 25 meter HP-1, 0.32 mm i.d., 0.17 μm film column (Hewlett-Packard) was used.

Supplies and Reagents

Tributyltin chloride and tripropyltin chloride (Aldrich Chemical Co., Milwaukee, WI) standards were prepared in HPLC grade dichloromethane (Fisher Scientific, Fair Lawn, NJ). Sodium borohydride was used as obtained (Fisher Scientific). Organosilicon standards (Petrarch Systems Inc., Bristol, PA) and diethylselenide standards (Strem Chemicals Inc., Danvers, MA) were prepared in HPLC grade toluene (Fisher Scientific). Solid phase extractions were performed using C18 SEP PAKs (Millipore Corporation, Milford, MA). Vitamin tablets containing 25 mg selenium as sodium selenate were obtained from a local retailer and o-phenylenediamine was obtained from Eastman-Kodak, Rochester, NY.

Results and Discussion

Concentric Dual Flow Torch (CDFT). This design (Figure 1) utilizes two concentric alumina tubes and two plasma gas flows. The first gas flow passes between the two tubes while the second flow passes through the inner tube. The two tubes are held concentric by a thin wire wrapped around the inner tube. The gas flow passing between the tubes flows around this wire in a helical path and out along the walls of the outer tube. This arrangement holds the plasma away from the walls and makes it more spatially stable. Plasma centering in the discharge tube reduces interactions with the walls. The second make-up flow, passing through the inner tube, is usually a relatively low flow to sweep analyte into the plasma at a relatively low rate, hence providing long residence times of analytes in the plasma. The tubes are held in place by brass fittings that thread directly into the microwave cavity. Plasma gas flows are introduced via Swagelok fittings that thread into the walls of the brass fittings. Reagent gases are added to the plasma gas before the attachment to the torch. Both plasma flows can be controlled independently.

This arrangement allows splitless injection of up to 1μl without significant disruption of the plasma and provides the sensitivities and selectivities that are obtainable with conventional torches. The combination of dual plasma gas flows and the alumina discharge tubes allows the operation of the plasma over a wider range of conditions and results in reduced baseline drift. Alumina tubes are also less expensive than quartz tubes.

Solvent Capacity. The most notable feature of the CDFT is the large solvent capacity. The ability to handle splitless injection of up to 1 µl greatly expands the utility of AED, especially if dilute solutions must be analyzed. The capacity for splitless injection of eight different solvents has been determined and Table I shows that several commonly used solvents can be injected in the splitless mode. It was found that as the carbon content of the solvent increased, the ability of the plasma to handle the solvent decreased. The plasma would accept 0.5-5 µg/s of carbon depending on the solvent used.

Table I.   Capacity for Splitless Injection of 8 Different Solvents

Solvent	50W	75W
Carbon Tetrachloride	1µl	1µl
Trichloromethane	1µl	1µl
Dichloromethane	1µl	1µl
Methanol	1µl	1µl
Diethylether	0.20µl	0.50µl
Petroleum Ether	0.20µl	0.25µl
Hexane	0.10µl	0.20µl
Toluene	0.10µl split 10:1	0.10µl split 10:1

Plasma Background and Alumina Tubes. The background spectrum from 200-500 nm obtained with the CDFT shows most of the emission lines expected for a helium MIP. However, the alumina discharge tubes give no silicon spectral background and two silicon emission lines (251.6 nm and 288.1 nm), which are very prominent when quartz discharge tubes are used, were not present. This allows improved detection of silicon and makes available an area of the spectrum for the monitoring of other emission lines that would have been obscured by silicon emission. The background spectrum showed no emission from aluminum at its most intense emission lines in this region (394.4 nm and 396.1 nm).

Reduced drift observed is considered to be a result of the outer plasma flow and the alumina tubes. As noted above, the outer flow holds the plasma more spatially stable and reduces the deposition of carbon soot. This flow along the walls also cools the discharge tube, and this, in combination with the thermal properties of the alumina, reduces discharge tube erosion. Because the inner walls of the discharge tube are not changing and the plasma is more spatially stable, less baseline drift occurs.

The superior thermal properties of alumina over quartz extend discharge tube lifetime (18). The higher melting point of the alumina allows the operation of the plasma at relatively high powers and low make-up flows without melting the discharge tubes. The ability to sustain high power plasmas combined with the versatility of two independent plasma gas flows, allows fine tuning of plasma operating conditions to enhance sensitivity and selectivity for specific elements.

Optimum Plasma Conditions, Limits of Detection and Selectivity. Table II lists the optimum CDFT flow and power combinations for carbon, tin and selenium. For both tin and selenium 1-2 ml/min of hydrogen was added to the inner plasma gas flow. Table III lists their limits of detection. The results obtained with the concentric dual flow torch are compared with those obtained with a conventional torch design (5). The CDFT was shown to be as sensitive and selective for these elements as the conventional torch.

The response factors for different elements have been shown to be a function of total plasma gas flow rate and two general types of reponses have been demonstrated (5). Non-metals such as selenium, respond best at relatively low flows. Their responses decrease rapidly with an increase in flow beyond the optimum and decreases slowly when the flow is decreased.

Table II.	CDFT Optimum Operating Conditions for Carbon, Tin, and Selenium Detection		
	C(193 nm)	Sn(284 nm)	Se (196 nm)
Inner Flow (ml/min)	40	200	60
Outer Flow (ml/min)	160	330	10
Power (W)	60	55	55

Metallic elements such as tin give the best response at higher plasma gas flows and the response drops off rapidly if the flow is increased or decreased from the optimum. These two types of behavior were seen for selenium and tin with the CDFT. In the case of tin detection at 284 nm, selectivity over carbon also increased dramatically as the flow was increased to give maximum sensitivity.

Supercritical Fluid Chromatography-MIP (SFC-MIP).   The MIP is an excellent gas chromatographic detector but some of the limitations described above are major problems when it is coupled to a SFC system The greatest problem results from the low solvent capacity of the MIP. In the GC mode only small amounts of helium are added to the plasma from the column, but the SFC column adds relatively large amounts (2-10 ml/min) of molecular species (usually CO_2). These large amounts are difficult for the plasma to atomize, and maintenance of a stable plasma in conventional torches becomes very difficult. If a plasma can be maintained, the large amounts of CO_2 entering it cause high levels of broad band molecular emission.

Many emission lines commonly monitored in the GC mode can then no longer be differentiated from the background. The large amounts of molecular species entering the plasma absorb much of its power; this leaves less energy available for atomization and excitation of analyte species, and results in reduced sensitivity.

Another problem occurs when density programming is used. Density programming in SFC is analogous to temperature programming in GC or gradient programming in liquid chromatography (LC). It results in a varying eluent flow rate and an increasing volume of eluent reaching the plasma per unit time. Changes in spectral background occur as the density of the mobile phase is increased and more eluent enters the plasma. The increasing eluent flow also reduces sensitivity, thus making calibration much more difficult.

The large sample capacity of the CDFT makes this design an excellent candidate for use as an AED for SFC. Stable plasmas were maintained under a wide range of conditions, various combinations of make-up flows and plasma powers being possible. With CO_2 from a capillary SFC system at 100 atm. entering the plasma, increased background levels (compared to a pure helium plasma) were seen throughout the spectrum especially at 248 nm (C), 400 nm (CN) and 470 nm (C_2). The region above 400 nm remains relatively free from molecular emission. Despite the more complicated background, detection of silicon at 251.6 nm was still possible. Figure 2 shows the SFC separation of two silicon containing compounds, each peak containing approximately 4 ng of silicon.

Table III.			GC-AED of Carbon, Tin and Selenium	
	LOD	LDR	Selectivity (C)	Selectivity (S)
Concentric				
Dual Flow Torch				
C 193 nm	20 pg	10^3	----	----
Sn 284 nm	20 pg	10^3	>30K	----
Se 196 nm	55 pg	$<10^3$	>50K	36K
Conventional Torch				
C 248 nm	12 pg	10^3	----	----
Sn 284 nm	6 pg	10^3	>30K	----
Se 204 nm	62 pg	10^3	10K	100
HP-5921A				
Cooled Torch				
C 193 nm	2.0 pg	10^3	----	----
Sn 303 nm	0.5 pg	10^3	>30K	----
Se 196 nm	4.0 pg	10^3	>50K	----

LOD- Limit of detection 3 times the signal to noise ratio
LDR- Linear dynamic range

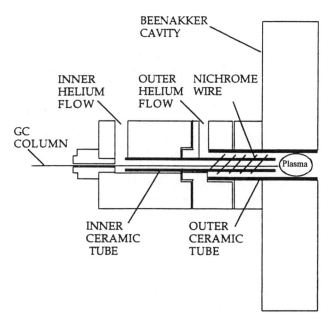

Figure 1.   Concentric dual flow torch (CDFT).

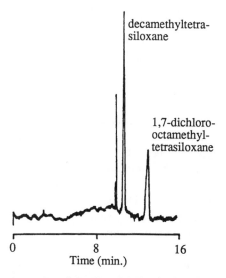

Figure 2.   SFC separation (80 atm, 50°C) of silicon containing species with AED of Si at 251.6 nm using the CDFT. Inner and outer plasma gas flows 160 ml/min, 70W.

Element Specific GC Dectection

Alkyltin Speciation in Sea Water. Organotin compounds have become a substantial environmental problem (19) since the use of alkyltins as herbicides, pesticides and as antifoulants in boat paints has led to the environmental appearance and persistence of alkyltin species. Tributyltin (TBT) chloride can be harmful to shellfish even at the part per trillion level (19).

Many methods of analysis for alkyltin speciation have been developed (20-27). These include pre-column hydride generation in aqueous solution coupled with purge and trap techniques, post column hydride generation or alkylation by Grignard reagents, as well as GC and LC separations of the alkyltin chlorides directly. Element selective detection methods include flame photometry, atomic absorption spectroscopy and direct current plasma atomic emission spectroscopy.

On-line Hydride Formation. Conversion of the alkyltin chlorides to hydrides has several advantages. Alkyltin chlorides are difficult to gas chromatograph directly because they interact with active sites in the chromatographic system, causing poor recovery of injected material and tailing peaks. Converting the chlorides to hydrides lowers their boiling points, increases stability, improves peak shape and increases sensitivity. Hydride generation provides another dimension of selectivity in addition to the chromatographic selectivity provided by the column and the spectroscopic selectivity provided by the spectrometer. On-line hydride generation provides a chemical selectivity which can be used to help identify analytes of interest.

Craig *et al.* have developed an on-line hydride generation technique (28,29), carried out by placing a small amount of sodium borohydride (NaBH$_4$) inside the injection port and injecting extracted tin chlorides through it. Hydrides form in the injection port and are then separated by the GC column. This method was shown to be quick, easy and free of sample contamination because no glassware or mixing of reagent solutions was required.

Solid Phase Extraction. Extraction techniques include purge and trap after hydride generation, liquid/liquid extraction and solid phase extraction (SPE) (23, 24, 25). The latter is appealing because it requires no reagent solutions or tedious separation of phases and may be automated more easily. As with the on-line hydride generation technique, less glassware is involved thereby reducing the chances for sample contamination or losses. The combination of SPE and on-line hydride generation with GC-MIP gives a fast, simple, sensitive and extremely selective method capable of determining part per trillion levels of alkyltin species in sea water.

CDFT. On-line hydrides have been generated from tripropyltin (TPT) chloride and tributyltin (TBT) chloride, the optimum injection port temperature being 250-26°C. This range gave complete conversion to the hydrides without redistribution reactions between the various alkyltin species present.

Figure 3 shows split and splitless 0.5 μl injections of the same standard containing TPT and TBT chlorides, using on-line hydride generation. The injection port temperature was 260°C. It can be seen that chromatographic integrity is maintained and complete conversion of the chlorides is achieved in the splitless mode. The ability to inject samples in the splitless mode without disruption of the plasma allows the direct analysis of ppb solutions of TBT.

HP-5921A AED System.   Limits of detection and selectivities for tin and selenium are listed in Table III. Solid phase extraction of 500 ml samples of sea water was performed by pulling the samples through a SEP PAK under a slight vaccuum. Extracted components were eluted with 10-20 ml of dichloromethane and concentrated by evaporation to 1 ml; splitless 5 μl injections of the extracts were made. The extraction of TBT from spiked water samples and unspiked sea water samples was successful. Chromatograms of a 1 ppb standard and water samples from Newport, RI. and New London, CT. are shown in Figures 4 and 5. In addition to the TBT observed, several contaminants that appear to contain tin were also seen. After examining each step of the sample preparation procedure it was concluded that the contaminants probably desorbed from the SEP PAK tube, where they may have been used as plastic stabilizers. It was found that pre-washing the SEP PAK with 10-15 ml of dichloromethane before the extraction of samples removed these contaminants and they no longer appeared in the sample extract. Further investigation into the source and identity of these contaminants is in progress. Future work will also include validation of this method using SPE, on-line hydride generation and GC-AED.

Selenium Dectection

CDFT.   Selenium is found in many different samples ranging from fuel oils, to plants, to vitamin tablets. Due to the similar chemistries of selenium and sulfur, many samples that contain selenium also contain sulfur. Therefore, in order to determine which components contain selenium, it is necessary to have good selectivity over not only carbon but also sulfur. Figure 6 illustrates the selectivity for selenium over carbon and sulfur. This sample contains 560 pg of selenium as diethylselenide, 170 ng of carbon as tridecane and 430 ng of sulfur as isobutylsulfide. The selectivity was 50,000 over carbon as tridecane and 36,000 over sulfur as isobutylsulfide, these results being summarized in Table III.

HP-5921A AED System.   Figure 7 shows the carbon and selenium chromatograms for a vitamin tablet extract. The tablet was digested in concentrated nitric acid and then reacted with o-phenylenediamine to give the piaselenol, which was then extracted with three 5 ml aliquots of toluene. These were combined and evaporated to 2 ml and injected. The peak area indicated that the extract contained 10 ppm of selenium and represents a recovery of selenium from the tablet of ca. 80%.

It would be difficult to determine which peak was the analyte of interest from the carbon chromatogram alone, even if a standard were available, but the selenium

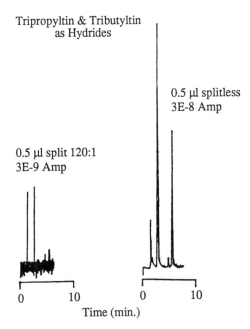

Figure 3. Split and Splitless 0.5μl injections of organotin species with on-line hydride generation and AED of Sn at 284 nm using the CDFT. Oven 180° isothermal.

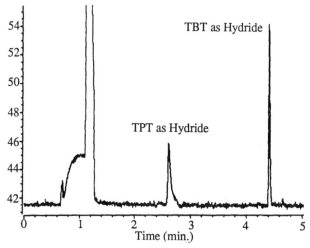

Figure 4. GC-AED (Sn 303 nm) of a spiked water sample, spiked at 1 ppb each of TBT and TPT chloride, concentrated 100 times, 1μl splitless. Oven 80°C 1 min, 20 deg/min.

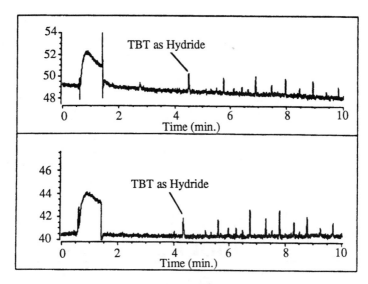

Figure 5. GC-AED (Sn 303 nm) of seawater extracts, concentrated 500 times, 5 μl splitless injection. Temperature program as in Fig 4.

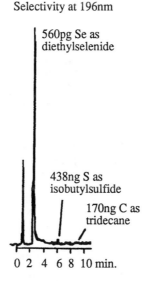

Figure 6. GC-AED of Se at 196 nm using the CDFT illustrating the selectivity over carbon and sulfur.

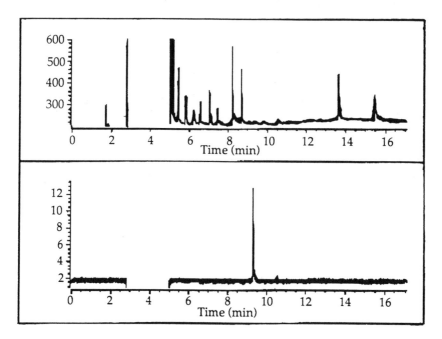

Figure 7.   GC-AED (C 193 nm (top) and Se 196 nm (bottom)) of a
vitamin tablet extract. 1 µl injection of toluene extract containing the
piaselenol derivative. Oven 40°C 1 min, 10 deg/min.

chromatogram shows only one very obvious peak that contains selenium and corresponds to the piaselenol. The sensitivity and selectivity of the 5921A greatly simplifies the interpretation of this chromatogram.

Conclusions

The unique features of the CDFT allow this design to be applied to a wider range of chromatographic problems than conventional torches. The large sample capacity of this torch allows the analysis of extremely dilute solutions and makes it possible for it to be used in capillary SFC-AED. The HP-5921A system has brought GC-AED to a high level of general utility and defines standards for future GC-AED systems. Development of applications such as alkyltin speciation in sea water, illustrate the power of these systems to detect selectively low levels of specific analytes in complex matrices.

The power and versatility of the MIP for chromatographic detection have yet to be fully realized. Through the development of improved instrumentation such as the CDFT and the HP-5921A AED system, combined with a wider range of established applications, the MIP will be able to meet the continuing demands for chromatographic detectors that provide specific information about eluents at extremely low levels.

Acknowledgments

We thank Bruce Quimby, James Sullivan and the Hewlett-Packard Company for their interest and support.. This work was supported in part by Merck, Sharp and Dohme Research Laboratories, Baxter Healthcare Corporation and the 3M Company.

Literature Cited

(1)   Beenakker, C. I. M. *Spectrochim. Acta* **1977**, *32B*, 173.
(2)   Quimby, B. D.; Delaney, M. F.; Uden, P. C.; Barnes, R. M. *Anal.Chem.* **979**, *51(7)*, 875.
(3)   Quimby, B. D.; Delaney, M. F.; Uden, P. C.; Barnes, R. M. *Anal.Chem.* **1980**, *52*, 259.
(4)   Carnahan, J. W.; Mulligan, K. J.; Caruso, J. A. *Anal. Chim. Acta* **1981**, *130*, 227.
(5)   Estes, S. A. ; Uden, P. C.; Barnes, R.M. *Anal. Chem.* **1981**, *53*, 1829.
(6)   Mulligan, K. J.; Caruso, J. A.; Fricke, F. L. *Analyst* (London) **1980**, *105*, 160.
(7)   Wasik, S. P.; Schwartz, P. *J. Chromatog. Sci.* **1980**, *18*, 660.
(8)   Estes, S. A.; Uden, P. C.; Barnes, R. M. *Anal. Chem.* **1981**, *53*, 1336.
(9)   Sullivan, J. J.; Quimby, B.D. *J. High Res. Chromatogr.* **1989**, *12*, 282.
(10) Goode, S. R.; Chambers, B.; Buddin, N. P. *Spectrochim. Acta* **1985**, *40B(1/2)*, 329.

(11)   Bollo-Karmara; Codding, E. G., *Spectrochim. Acta* **1981,** *36B*, (10), 973.

(12)   Feuerbacher, H.; Oppermann, M. "Improvements of the He-MIP by Discharge Tubes Made of Selected Ceramics" Poster #Th31, 1988 Winter Conference on Plasma Spectrochemistry, San Diego, CA.

(13)   Quimby, B. D.; Sullivan, J. J. *Anal. Chem.* **1990,** *62*, 1027.

(14)   Sullivan, J. J.; Quimby, B. D. *Anal. Chem.* **1990,** *62*, 1034.

(15)   Sullivan, J. J.; Quimby, B. D. *J. High Res. Chromatog.* **1989,** *12*, 813.

(16)   Quimby, B. D.; Delaney, M. F.; Uden, P. C.; Barnes, R. M., *Anal. Chem.* **1979,** *51*, 875.

(17)   Forbes, K. A.; Vecchiarelli, J. F.; Uden, P.C.,Barnes, R. M., *Anal. Chem.* **1990,**.*62,* 2033

(18)   Camman, K.; Lendero, L.; Feuerbacher, H.; Ballschmiter, K. *Fresenius Z. Anal. Chem.* **1983,** *316*, 194.

(19)   Seligman, P. F.; Grovhoug, J.G.; Valkirs, A. O.; Stang, P. M.; Fransham. R; Stallard, M. O.; Davidson, B.; Lee, R. F. *Appl. Organomet. Chem.*, **1989,** *3*, 31.

(20)   Carter, R. J.; Juroczy, N. J.; Bond, A. M. *Env. Sci. and Tech.* **1989,** *23*, 615.

(21)   Robbins, W. B.; Caruso, J. A.; Fricke, F. L. *Analyst* **1979,** *104*, 35.

(22)   Thorburn,D.; Glocking, F.; Hamot, M. *Analyst* **1981,** *106*, 921.

(23)   Randall, L.; Donard, O.; Weber, J. H. *Anal. Chim. Acta* **1986,***184*, 197.

(24)   Sullivan, J.J.; Torkelson, J.D.; Wekell, M. M.; Hollingworth, T. A.; Saxton. W. L.; Miller, G. A.; Panaro,K. W.; Uhler, A. D. *Anal. Chem.* **1988,** *60*, 627.

(25)   Muller, M. D. *Anal. Chem.* **1987,** *59*, 617.

(26)   Stallard M. O.; Cola, S. Y.; Dooley, C. A. *Appl. Organomet. Chem.* **1989,** *3 (1)* 105.

(27)   Krull, I. S.; Panaro, K. W.; Noonan, J.; Erickson, D. *Appl. Organomet. Chem.* **1989,** *3*, 295.

(28)   Clark, S.; Ashby, H.; Craig, P. J. *Analyst* **1987,** *112*, 1781.

(29)   Clark, S.; Craig, P. J. *Appl. Organomet. Chemi* **1988,** *2*, 33.

RECEIVED July 8, 1991

Chapter 6

Optical-System Developments for Plasma Emission Detection in High-Resolution Gas Chromatography

Karl Cammann, Michael Faust, and Karl Hübner

Lehrstuhl für Analytische Chemie, Anorganisch-Chemisches Institut der Westfälischen Wilhelms Universität, Wilhelm-Klemm Strasse 8, 4400 Münster, Germany

A short survey is given of the developments of optical systems with respect to element selectivity in high resolution gas chromatography coupled with plasma emission detection. In addition to improvements for a monochromator system, the Plasma Emission Detector (PED) based on oscillating interference filters is presented. Results are given for C, H, F, Cl, Br and S containing compounds with PED detection. Using an Atomic Emission Detector (AED), volatile halogenated compounds were investigated. Detection limits between 140-350 pg abs. and dynamic ranges over 4 orders of magnitude were found for C,H,Cl and Br.

The combination of atomic emission spectrometry and high resolution gas chromatography has been under investigation in this research group since 1981, when the GC - MIP - monochromator arrangement shown in figure 1 (1) was developed.

The helium plasma was generated in a Beenakker - type - cavity with the modifications of van Dalen et al. (2), and a self - built coupling device at atmospheric pressure. To reduce the electronic noise and to differentiate between photo - and dark current an AC modulated MIP was used in combination with the lock in amplifier technique. The He - MIP image was focussed on one end of a UV fibre optic to increase the flexibility of the whole system. By using ceramic (60 % Al_2O_3, 36 % SiO_2, 3 % Alkali) instead of quartz discharge tubes the need for solvent venting could be eliminated, because the normal amounts of solvent (1 - 2 μl) did not extinguish the plasma.

The detection limits for carbon, chlorine and bromine were 200/8, 200/13 and 400/26 pg/pgs^{-1}. The selectivities of Cl and Br with respect to C were about 10^4 and the linear dynamic range extended over 3 - 4 decades. A comparison to ECD detection is shown in figure 2 (1) for a synthetic sample of polychlorinated biphenyls (Chlophen A 60).

0097–6156/92/0479–0105$06.00/0

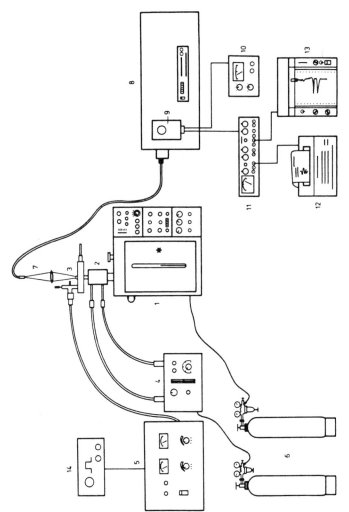

Figure 1: Experimental system for element-specific detection. 1 gas chromatograph; 2 GC-MIP coupling device; 3 microwave cavity; 4 control unit for coupling device; 5 microwave generator; 6 gas supply for GC and MIP; 7 optical system and light guide; 8 monochromator; 9 photomultiplier; 10 power supply for 9; 11 lock-in amplifier; 12 computing integrator; 13 recorder; 14 function generator. (Reproduced with permission from ref.1. Copyright 1983 K.Cammann)

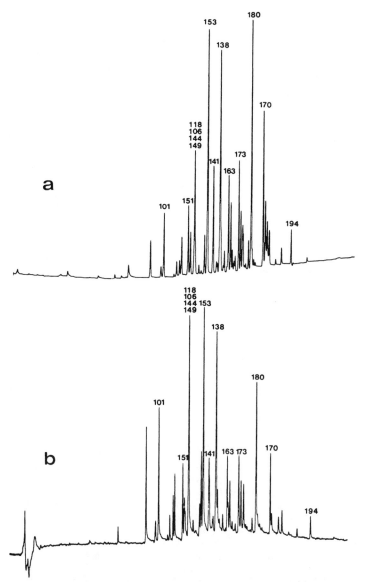

Figure 2: a: Chromatogram of Clophen A 60 with ECD-detection. Parameters of GC-separation: 2.2 ng A 60 in 1.2 µl n-hexane. Carrier gas: hydrogen; flow rate: 2 ml/min (40°C); temperature program: 5 min 40 °C, then heating from up to 140°C, heating rate 45°C/min, then heating from 140-250°C, heating rate: 3°C/min; column: fused silica capillary, liquid phase: SE 54, length: 25 m b: Chromatogram of Clophen A 60 with MIP-detection. Parameters of GC-separation and MIP-detection: 5.1 µg A 60 in 2.0 µl n-hexane. Carrier gas: helium; flow rate: 2 ml/min (40°C); temperature program: 5 min 40°C, then heating up to 140°C, heating rate: 45°C, then heating from 140-250°C, heating rate: 1°C/min; column: fused silica capillary, liquid phase: OV 1, length: 50 m. (The identification of the peaks was according to K. Ballschmiter (15)) (Reproduced with permission from ref.1. Copyright 1983 K.Cammann)

Instrumental Developments

Optimization of the Monochromator System. To improve the sensitivity of the foregoing system, efforts were then made to reduce the background noise (*3*). To reduce the detection of background light emission from the ceramic discharge tube an aluminium - plate with a 1.4 mm opening was placed 10 mm above the discharge tube. It was shown by adding small amounts of a gaseous compound to the helium make - up gas stream that this optical mask reduced the noise without decreasing the absolute light intensity.

To improve the stability of the signal and to reduce the dark current of the photomultiplier the monochromator was partially filled with argon and the photomultiplier was equipped with peltier cooling. Both techniques brought slight improvements to the noise level.

Background Correction. An oscillating quartz plate was mounted just before the exit slit of the monochromator to perform background correction. Although the sensitivity was about a factor of two worse by comparison with the power modulated plasma, the simultaneous detection of background and signal increased the element selectivity. This effect is shown in figure 3 (*3*). Not only are interfering signals supressed in a bromine selective chromatogram, but also interfering compounds produce negative peaks, in contrast to the bromine containing compounds. A satisfactory explanation for this effect has yet to be found.

A Plasma Emission Detector (PED) system using Oscillating Interference filters

The inflexibility of a monochromator gives some disadvantages in such an optical system. In addition to daily recalibration, the monochromator has to be kept at a constant temperature and vibrations avoided. Any deviation from the standard operating conditions may cause a change in the response and therefore decreases the reproducibility of the results.

To avoid these problems a simple optical system was constructed (*4,5*), the central feature of which consists of interference filters with narrow band width. Their transmission wavelength is tuned to the emission wavelength of the elements to be detected. It is well known that the exact center wavelength of an interference filter is shifted towards shorter wavelength, if the angle between the filter surface and the light incidence differs from 90 degrees. Thus a controlled oscillation of the filter scans in the two extreme positions the element emission signal on one side, and the background emission on the other. Combined with the phase and frequency selective lock in amplifiers a simultaneous background compensation is achieved. A schematic diagram of this filter oscillation system is given in figure 4 (*4*).

The detection limits obtained with the Plasma Emission Detector (PED) were 2 ng absolute for chlorine (measured at 479.5 nm), 10 - 15 ng abs. for bromine (470.5 nm) and 10 - 13 ng abs. for carbon (247.9 nm). Regarding the selectivity of the PED, the same effect that was observed with the background compensated monochromator. True element signals were detected in the positive direction and interfering element signals were detected in the negative direction. For chlorine and bromine containing compounds, the linear dynamic range covered three decades. Interelement ratios were also calculated for bromine and chlorine containing compounds, the results showing a relative standard deviation of 8 -11 %.

To investigate the assumption that PED detection is element selective and independent of molecular structure, different chlorinated and brominated substances were measured (*5*). The chlorine concentration covered a range of 5 - > 5000 ng absolute. By correlating the chlorine peak areas with the known chlorine masses, calibration plots were obtained with correlation coefficients of

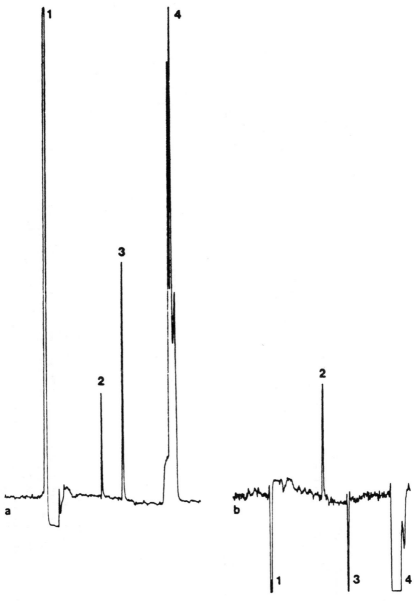

Figure 3: Specificity enhancement for bromine. Microwave energy: 65 W for figure a, 80 W for figure b. Emission line used: bromine 470.49 nm. Chromatogram "a" is obtained with power-modulated plasma, chromatogram "b" is obtained with continous power plasma and background-correction with an oscillating quartz-plate. Peak identification: 1 solvent (n-hexane), 2 1-bromo-3-chloropropane, 3 decane, 4 hexadecane (Reproduced with permission from ref.3. Copyright 1984 K.Cammann).

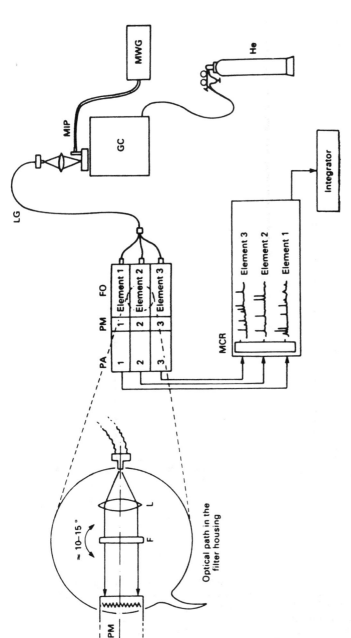

Figure 4: Schematic diagram of the GC-MIP filter-oscillation system: F oscillating interference filters; FO optical filter and filter oscillation driving element; GC gas chromatograph; He helium gas supply for GC and MIP; L lens; LG light guide optic; MIP microwave-induced plasma cavity; MWG microwave generator; PA photoamplifier (lock-in type); PM photomultiplier tube; MCR multi-channel recorder. (Reproduced with permission from ref.4. Copyright 1988 K.Cammann)

0.997 - 0.999. For the substances investigated this was a good indication of a molecular structure independent detection. Further studies dealt with the detection of other elements and the optimization of the chromatographic and optical system. Attempts to detect hydrogen and deuterium simultaneously failed, because the full width at half maximum of the interference filters used (656.12 nm and 657.05 nm) were too broad for an interference - free detection (*6*). For hydrogen (656.12 nm) and sulfur (545.5 nm) detection limits of 1.8 ng absolute and 9 ng abs. respectively with a dynamic range over three decades were found (*7*).

To ensure that the chosen oscillation frequency of 20 Hz and the signal processing frequency of the lock in amplifiers in the first order were correct, investigations with a Fast Fourier Transform Analyser were undertaken (*6,8,9*). Recording the frequency spectra in the range from 0 - 1000 Hz and using different concentrations of the analyte (50 - 1000 ng absolute of trichloromethane) gave this proof since the signal at 20 Hz dominated the frequency spectra of all investigated elements and concentrations. Figure 5 (*9*) shows a three dimensional plot of the carbon channel of trichloromethane (concentration 50 - 1000 ng absolute of trichloromethane).

The next subject of interest was the detection of the element fluorine. The results gave a detection limit of 5 ng absolute but a poor linear dynamic range of two decades (*8,9*). During the measurements of reproducibility it became obvious that high concentrations of fluorine containing compounds caused a steady signal decrease. To confirm this observation, calibration gases were used and a mixture of 0.011 vol % chlorotrifluoromethane in helium showed the same effect. By diluting the gas mixture with additive helium (1 : 10) the effect disappeared and a relative standard deviation of 0.58 % was achieved (*10*). The explanation for this effect was given by Scanning Electron Microscopy (SEM) pictures of the ceramic tube materials. The tubes used for high concentrations of the analyte showed strong corrosion of the inner surface (*11*), thus providing at least a partial reason for the effects with fluorine.

In another study, interelement ratios for hydrogen and carbon were calculated. Using four different organic compounds (dimethylsulfoxide, thiophene, toluene and n - heptane) the deviations from the theoretical values were about 1 - 7 % (*12*). Current studies of the Plasma Emission Detector deal with the determination of adsorbable organic compounds with a modified helium MIP and the detection of further elements like phosphorus, oxygen, nitrogen and iodine.

Comparison with a Photodiode Array Based Instrument

In connection with a research project dealing with the determination of volatile halogenated compounds in various matrices, a photodiodearray based Atomic Emission Detector (AED) (Hewlett Packard) was investigated. In contrast to the PED, the AED uses a diffraction grating as the wavelength dispersing element with a photodiode array as the multichannel optical sensor (*13*). Assuming that detection by the AED is independent of molecular structure, various volatile chlorinated hydrocarbons were determined. Calibration plots for the elements carbon, hydrogen and chlorine were obtained by correlating the resulting peak areas with the corresponding amounts (*14*). The calibration plots are shown in figures 6,7 and 8.

Considering the good correlation coefficients obtained, the conclusion is justified that for this group of substances response is independent of the molecular structure.

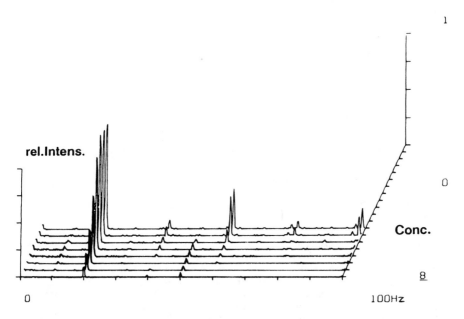

Figure 5: Three dimensional plot of the carbon channel of Trichloromethane.
(Reproduced with permission from ref.9. Copyright 1989 K.Cammann)

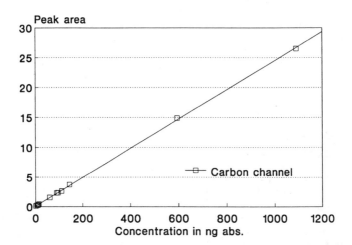

Figure 6: Calibration plot for the carbon-channel (496 nm). Substances:
Trichloromethane, Tetrachloromethane, 1,1,1-Trichloroethane, Trichloroethene
and Tetrachloroethene. Each value average of 15 injections. Correlation
coefficient : 0.9999.

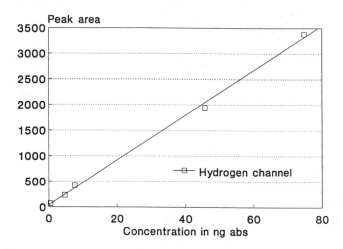

Figure 7: Calibration plot for the hydrogen-channel (489 nm). Substances: Trichloromethane, 1,1,1-Trichloroethane and Trichloroethene. Each value average of 15 injections. Correlation coefficient : 0.9989.

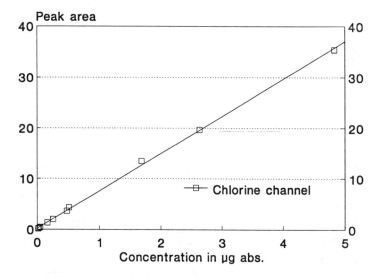

Figure 8: Calibration plot for the chlorine-channel (479 nm). Substances as under figure 6. Each value average of 15 injections. Correlation coefficient : 0.9996.

Volatile chlorinated hydrocarbons in blood as a biological matrix were also determined. Using headspace injection the samples were analysed with either an ECD or the AED (*14*). The ECD sensitivities exceeded those of the AED by factors of 2.5 - 2000 (see tables I and II), depending on the compound analysed, but the linear dynamic range of the AED reaches higher values.

Table I: Dynamic ranges and experimental detection limits of blood analysis by HS-GC-AED

	Dynamic range	lowest observed concentration	
		Cl-channel	C-channel
	(μg/l)	(μg/l)	
Dichloromethane	1704 - >62500	68	68
Trichloromethane	733 - >73300	73	365
Tetrachlormethane	790 - >78080	79	790
1,1,1-Trichloroethane	674 - >67000	67	135
Trichloroethene	731 - >73080	149	149
Tetrachloroethene	803 - >80270	773	803

Table II: Dynamic ranges and experimental detection limits of blood analysis by HS-GC-ECD

	Dynamic range	lowest observed concentration
	(μg/l)	(μg/l)
Dichloromethane	66.2 - > 17500	26.5
Trichloromethane	1.5 - > 3720	1.5
Tetrachloromethane	0.4 - > 1580	0.39
1,1,1-Trichloroethane	0.33 - > 3320	0.33
Trichloroethene	1,5 - > 3680	1.5
Tetrachloroethene	6.3 - > 4020	0.39

This property offers the opportunity to analyse strongly contaminated samples with much reduced calibration efforts. Currently interest lies in expanding the variety of detectable compounds with the AED. In Figure 9 a chromatogram of eight brominated and chlorinated compounds is shown with respect to the bromine and chlorine channels.

For the compounds trichloromethane, trichloroethane, tetrachlorome-thane, trichloroethene, monobromdichloromethane, dibrommonochlorome-thane, tetrachloroethene and tribromomethane, it appears that for all four detected elements the response is independent of the molecular structure. In Table III the linear dynamic ranges and the experimental detection limits are given for the four elements.

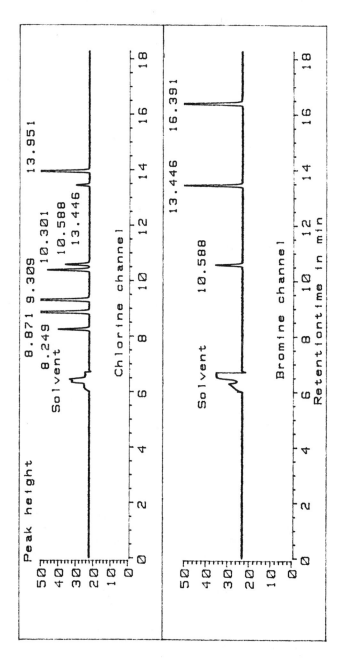

Figure 9: Chromatogram of the bromine and chlorine channel of eight different compounds. Concentrations are in the range of 20-50 ng of the respective compounds. Chromatographic parameters : Column SE 54 DF 50 m * 0.32 mm ID with film thickness 0.31 μm, Carrier gas helium 99.996% purity, Column flow 1 ml/min, Split ratio 50 : 1, Temperature program 10 min at 40 C, with 20 C/min to 250 C, Make up gas flow 40 ml/min, Cavity pressure 1.7 psi.

Table III: Linear Dynamic Ranges and Experimental Detection Limits of
Carbon, Hydrogen, Chlorine and Bromine Compounds

Element	Linear dynamic range	Experimental Detection Limit
	ng absolute	ng absolute
Carbon	0.70 - 1200	0.35
Hydrogen	0.42 - 620	0.14
Chlorine	0.21 - 4400	0.21
Bromine	0.34 - 2800	0.17

Detection limits need to be further improved, the main reasons for lack of
sensitivity being the lack of optimization of make up gas flows, and the
presence of impurities in the helium gas. After optimization the detection
limits of the detected elements should reach the low picogram range.

Acknowledgement

This work was kindly supported by the Bundesministerium für Forschung und
Technologie, Project Nr. 325 - 4007 - 07 INR 245 / 223

Literature Cited

(1) Cammann, K., Lendero, L., Feuerbacher, H., Ballschmiter, K. Fresenius Z
Anal. Chem. **1983**, 316, 194
(2) Van Dalen, J.P.J., de Lezenne Coulander, P.A., de Galan, L. Spectrochim.
Acta Part B **1978**, 33, 545
(3) Lendero, L., Cammann, K., Ballschmiter, K. Mikrochimica Acta **1984**, 1,
107
(4) Müller, H., Cammann, K. Journal of Analytical Atomic Spectrometry **1988**,
3, 907
(5) Cammann, K., Müller, H. Fresenius Z Anal. Chem. **1988**, 331, 336
(6) Faust, M., Diploma thesis, Münster, **1988**
(7) Faust, M., Stilkenböhmer, P., Cammann, K., 17th Int Symp on Chrom,
Vienna, **1988**, 2, 142
(8) Faust, M., Buscher, W., Cammann, K., Bradter, M., Winter, F., Europian
Winter Conference on Plasma Spectrochemistry, Reutte, **1989**, Abstract Nr. P
1-25
(9) Bradter, M., Buscher, W., Cammann, K., Faust, M., Winter, F. Michrochi-
mica Acta **1989**, 3-6, 215
(10) Bradter, M., Buscher, W., Faust, M., Quick, L., Winter, F., Cammann, K.
Fresenius Z Anal. Chem. **1989**, 334, 718
(11) Buscher, W., Diploma thesis, Münster, **1989**
(12) Stilkenböhmer, P., Cammann, K. Fresenius Z Anal. Chem. **1989**, 335, 764
(13) Sullivan, J.J., Quimby, B.D. J. High Resolution Chromatography **1989**, 12
(5), 282
(14) Cammann, K., Bradter, M., Buscher, W., Faust, M., Quick, L., Winter, F.,
Winter Conference on Plasma Spectrochemistry, St.Petersburg, Florida, **1990**,
Abstract Nr S4
(15) Ballschmiter, K., Zell, M. Fresenius Z Anal. Chem. **1980**, 302, 20

RECEIVED May 6, 1991

Chapter 7

Atomic Emission Detectors for Gas Chromatography
Twelve Years of Industrial Experience

John S. Marhevka, Donald F. Hagen, and Joel W. Miller

3M Corporate Research Laboratories, 3M Center, Building 201–1S–27, St. Paul, MN 55144–1000

Industrial applications of commercial atomic emission detectors at 3M are described. Specific examples illustrate the advantages of this type of detector: element specificity, particularly for the halogens, and compound-independent response.

The use of derivatizing reagents containing heteroelements provides a way of tagging specific functional groups for ready identification by monitoring the specific tagging element. Plasma pyrolysis experiments allow for the elemental characterization of non-volatile samples.

The 3M Corporate Research Laboratories acquired the first commercial microwave emission detector for gas chromatography in the United States in 1978. Approximately ten MPD 850 Microwave Plasma Detectors (Applied Chromatography Systems, Ltd., Luton, England) were sold in this country before production ceased due, in part, to distribution problems and a general lack of published applications. The purpose of this paper is to supplement some of the excellent reviews of the technique *(1-3)* with a summary of our industrial applications over the last twelve years, with the most recent work done on a current commercial instrument (H-P 5921A, Hewlett-Packard, Avondale, PA).

The MPD 850 System

As received from the manufacturer, the MPD 850 was originally configured for packed column gas chromatography, and no suitable means of data reduction was provided. Our instrument was rapidly converted to capillary column operation, and a dedicated Hewlett-Packard 3354 Data System was added *(4)*. The data system enabled us to monitor eight different elemental channels of the MPD simultaneously, time synchronized, with a sampling rate of 8 Hz. The desire to compare the

0097–6156/92/0479–0117$06.00/0

response of the MPD 850 elemental channels to conventional detectors, such as the FID and ECD, encouraged the installation of inlet splitting of the sample between two matched capillary columns. The 0.75 m spectrometer is thermostatted and has a movable primary slit allowing the element profiles to be reconfigured on secondary slits on the Rowland circle; generally, this is not necessary as the spectrometer is quite stable. Scavenge gases of oxygen or nitrogen are added to the helium plasma gas to reduce the carbon continuum and minimize carbon deposits in the discharge tube; the choice of the optimum scavenge gas is the principle limitation on the number of elements which can be monitored simultaneously by the direct reading spectrometer--a determination of both oxygen and nitrogen would require a change of scavenge gas and a second injection. The instrument was provided with a variable "ghost" correction which subtracts a portion of the carbon signal from the other elemental channels so that, for example, a large carbon response does not elicit a response on the chlorine channel. Unlike most of the devices described in the literature, and the current commercial offering, this unit utilizes a quarter-wave Evenson cavity operating at a reduced pressure of 1-5 torr.

Applications. Figure 1 illustrates one of our most effective applications of element selectivity: a human blood plasma sample (coded H215) taken following an industrial exposure to a fluorocarbon. The nonspecific flame ionization detector response illustrates the complexity of the chromatogram typical of biological samples when using conventional or mass spectral (total ion) detectors. The electron capture detector is somewhat more selective, but there are present a considerable number of electrophilic compounds of unknown structure and response. In contrast, the fluorine channel of the MPD is unequivocal in identifying the single fluorine-containing species in the sample, the methyl ester of perfluorooctane carboxylic acid; it would also be useful in pinpointing the region of the chromatogram to examine with GC/MS for a more conclusive structural verification. An attempt to use single ion mass spectrometry was unsuccessful; excessive fragmentation of this compound provided no suitable ion that could be distinguished above the hydrocarbon interferences present in this sample. In this case, where the analyte was known to be a compound with a perfluorinated carbon chain, it is possible to calibrate for ng levels of fluorine with the MPD and achieve a very good approximation of the level of unknown components (metabolites which might be observed at various times after exposure) in the absence of pure standards.

A metabolic study was undertaken in which rats were fed 1H,1H,2H,2H-perfluorodecanol and sacrificed at various times post dose. Blood and urine samples were prepared for analysis by ECD and MPD *(5)*. Rat No. 12 (Figure 2) was sacrificed two hours after feeding. The original alcohol (peak no. 3) is virtually absent in the blood plasma after this brief metabolism time. Rat No. 13, which was sacrificed twelve hours after feeding, gave a plasma sample containing the final metabolite, the derivatized perfluorooctanoic acid (peak no. 1) as the major component. The other metabolites are intermediates to the final metabolite, and their concentrations vary with the time allowed for metabolism. Retention time

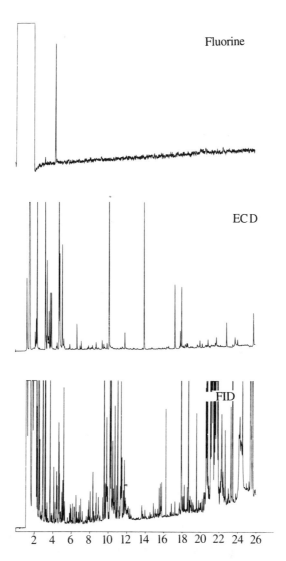

Figure 1. Chromatograms for sample H-215, human blood plasma. Detector response is plotted vs. elution time (minutes) for the fluorine, electron capture, and flame ionization detector response. The peak on the fluorine channel is normalized to give a full-scale signal. (Reproduced with permission from ref. 4. Copyright 1983 Pergamon Press, Inc.)

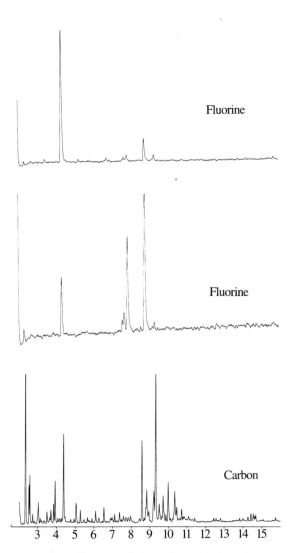

Figure 2. Chromatograms from rats fed 1H,1H,2H,2H-perfluorodecanol. Top chromatogram represents fluorine response for rat 13; middle trace represents fluorine response for rat 12, and bottom trace is the carbon response for rat 12.

matching against known commercially available or synthesized standards and fluorine NMR suggest that two of the intermediates after derivatization are 2H,2H-perfluorodecanoate (peak no. 4) and its corresponding unsaturate (peak no. 2). The remaining major metabolite is unidentified but contains an esterifiable functional group. The presence of high inorganic fluoride levels in the rat plasma and the presence of perfluorooctanoate confirm the defluorination of the starting alcohol. This study indicates that, a rat population exposed to either 1H,1H,2H,2H-perfluorodecanol or perfluorooctanoic acid would exhibit the same component in plasma, one resulting from metabolism and one from direct exposure. Unless a metabolic study were undertaken, the source of the exposure would remain in question.

Elemental Derivatization. The element specificity of the microwave emission detector can be enhanced through the use of derivatization reactions in which a derivatizable group is reacted with a reagent containing a unique or different element, or combination of elements, which can then be monitored by this detector. This elemental tag adds additional specificity, thus facilitating identification of those functional groups which have reacted with the derivatizing agent. Several of the common derivatization reactions, such as esterification, acetylation, and silylation have been examined in which specialized reagents were utilized to tag common fuctional groups *(6)*. For example, any halogenated alcohol can be used with BF_3 etherate to form the halogenated ester of a fatty acid (in our experiments, we used trifluoroethanol to form the fluorine-labeled esters of the C_{10}-C_{18} even-numbered fatty acids). Alternately, a suitably-tagged acetylating reagent can be used to derivatize a series of alcohols--we used the commercially-available chlorodifluoroacetic anhydride to tag the C_6-C_{12} even-numbered alcohols with both chlorine and fluorine. Similarly, the same reagent was used to acetylate a series of aliphatic amines, as shown in Figure 3. Since the starting mixture contained equivalent amounts of each amine by weight, the carbon area response is approximately the same across the series, while the chlorine and fluorine responses follow the expected decrease as the aliphatic amine carbon chain length increases relative to the single derivatized group present in each homolog. A plot of the peak area ratio for various elemental channels versus the total carbon number for these derivatives is shown in Figure 4. The lines exhibit slopes related to a homologous series: the Cl/C and F/C ratios decrease as the length of the carbon chain increases, while the hydrogen remains proportional to the carbon content, and the F/Cl ratios remain virtually constant as the alkyl chain length increases. The fact that all of these plots are linear suggests that it is possible to predict where a member of the series would fall. A change in F/Cl ratio would indicate that there already is a halogen present in the molecule, prior to derivatization.

A mixture of four amino acids was acetylated with chlorodifluoroacetic anhydride and then esterified with diazomethane. The chromatogram of the derivatives is shown in Figure 5 with an elution order of leucine, methionine, phenylalanine, and lysine. The sulfur-containing methionine is readily identifiable

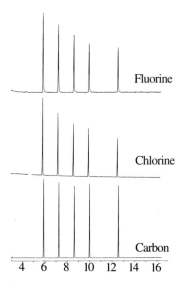

Figure 3. Normalized chromatograms of the chlorodifluoroacetic anhydride derivatives of pentyl, hexyl, heptyl, octyl, and decyl amines. (Reproduced with permission from ref. 6. Copyright 1985 Pergamon Press, Inc.)

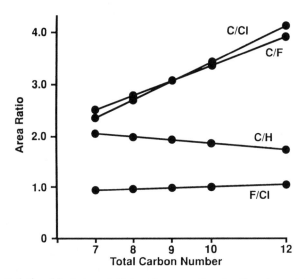

Figure 4. Relationship between the peak area ratios for the elemental channels and the total carbon number for the derivatized amines from Fig. 3. (Reproduced with permission from ref. 6. Copyright 1985 Pergamon Press, Inc.)

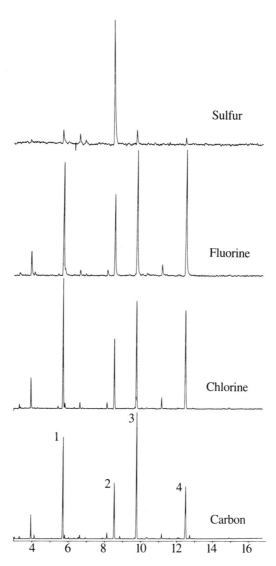

Figure 5. Sulfur, fluorine, chlorine, and carbon channel responses to the derivatized amino acids leucine (1), methionine (2), phenylalanine (3) and lysine (4). (Reproduced with permission from ref. 6. Copyright 1985 Pergamon Press, Inc.)

despite the small amount of ghosting (carbon molecular continuum response) observed on the sulfur channel. Lysine is easily distinguished in that it has two amino groups and gives twice the fluorine and chlorine response relative to carbon as leucine (peak 1). The relative C/H response (not shown) would also be highly suggestive of the presence of aromaticity in phenylalanine. An additional derivatization trial of phenylalanine was illustrative of the use of the element-specific channels of the MPD: in this case phenylalanine was acetylated as before, followed by alkylation with deuterated diazomethane. Two major peaks of roughly equal height were observed on the carbon, fluorine, and chlorine channels of the MPD, but only a single peak, the later-eluting of the pair, was observed on the deuterium channel. This suggests that a side reaction has occurred--the amino group is derivatized in both peaks, but only one of the pair has an active hydrogen which is susceptible to deuteroalkylation. One conclusion is that the earlier peak has been decarboxylated in a side reaction.

The sensitivity is about tenfold better for deuterium compared to hydrogen on the MPD 850, since one member of two different line pairs of different intensity is used for each element (hydrogen I (486.13 nm) and deuterium I (656.10 nm)). This fact suggested that deuterium derivatization or deutero-reduction would be worthwhile. Several commercially available reagents were examined: D_9-bis-trimethylsilyl acetamide for phenols, D_3 acetyl chloride for alcohols, and Deuter-8 for fatty acids. Deuterated diazomethane prepared from Aldrich starting materials resulted in the cleanest chromatogram (minimal impurities). Derivatizing reagents, such as $LiAlH_4$ or $LiAlD_4$ are often overlooked; $LiAlD_4$ was used to reduce the carbonyl group in acids, esters, aldehydes, etc. The deutero-reduction of benzonitrile and aniline with $LiAlD_4$ could proceed to form $C_6H_5CD_2ND_2$ from the nitrile and possibly $C_6H_5ND_2$, depending on the amount of deuterium/hydrogen exchange with the active hydrogens on aniline. Chlorodifluoroacetylation of these products was found to yield two peaks containing both chlorine and fluorine, but only one containing deuterium, suggesting that, under the conditions of this reaction, aniline does not undergo deuterium exchange.

Catalytic gas phase deuteration reactions were explored using the apparatus described previously (7). A sample could be injected into a normal injection port (unreacted), or passed through a short stainless steel tube packed with various catalytic materials, ranging from palladium or platinum Beroza catalysts (8) to a proprietary 3M nickel catalyst. The catalytic tube contained deuterium gas at 200 deg C; reaction times were normally 0.5 min. At the end of that time, the sample could be valved onto the capillary column in a split injection mode, thus eliminating nearly all of the excess deuterium. A sample containing decane, dodecene, dodecane, and tetradecane was analyzed as shown in Figure 6. The lower chromatogram is the MPD carbon channel response when the sample is injected into the normal GC injection port (i.e. not subjected to reaction with deuterium). The middle chromatogram represents the MPD carbon channel response when the sample passes through the reactor loop and the upper trace represents the deuterium channel response for the reacted sample. The alkene has been deuterated using the nickel

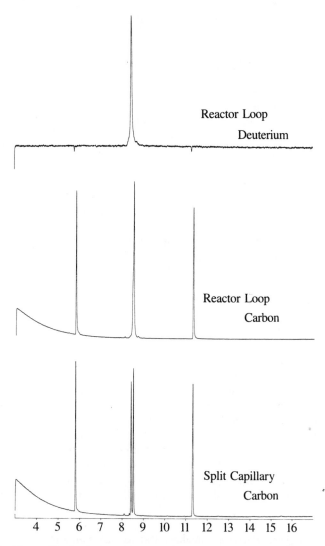

Figure 6. Normalized chromatograms of decane, dodecene, dodecane, and tetradecane with and without gas phase deuteration with a nickel catalyst. (Reproduced with permission from ref. 7. Copyright 1987 Pergamon Press, Inc.)

catalyst; the product peak coelutes with the C_{12} alkane, but now contains deuterium. Other catalysts are not as selective--in several cases, alkanes have been found to react to form deuterium-containing species. It is suspected that some isomerization and scrambling is occurring, i.e. a methyl group is formed, resulting in a family of branched chain isomers giving an asymmetric composite peak. This technique represents an excellent means of studying catalytic reactions.

Direct Plasma Pyrolysis. One final series of experiments with the Evenson cavity may not be possible on smaller plasmas. It is possible to insert a bare fused silica probe into the plasma discharge tube; the probe contains the residue of a solution of a nonvolatile material resulting from application of the solution to the probe and removal of the solvent with a heat lamp. Following probe insertion, the pressure in the microwave cavity is brought to 5 torr, the plasma is ignited, and the baseline is monitored using the appropriate channels of the MPD. At this point, the probe containing the sample is above the plasma. The applied power is then increased so that the plasma expands to engulf the probe, pyrolyzing the sample and generating a signal for the elements in the nonvolatile material. The response of the MPD for polyvinyl chloride is shown in Figure 7. Although there is a baseline shift during the process, particularly in the hydrogen channel, it does not preclude qualitative results. Other samples, such as sodium polyvinyltoluene sulfonate and some fluorocarbon elastomers, have been successfully examined by this technique, which is the subject of a U.S. patent *(9)*.

The H-P 5921A Atomic Emission Detector

The 3M Corporate Research Laboratories purchased the Hewlett-Packard 5921A in 1989. A complete description of the instrument is given elsewhere in this volume. Several differences and improvements over the MPD 850 should be noted here. A modified Beenaker cavity is utilized, with a water-cooled plasma which helps eliminate chemistry occurring at the wall of the discharge tube. The helium plasma operates at atmospheric pressure, and a series of scavenge gases are used to optimize response and selectivity for different elements. Sensitivity for oxygen and nitrogen is markedly improved such that trace level determinations of these elements is now routine. The scanning portion of the diode array detector covers a fairly narrow wavelength range, so for multielement determinations, it is often necessary to make multiple injections. Control by a data system with an automatic sampler maintains high precision and optimizes the system, including control of the various scavenge gases, for multielement analyses.

Applications. Deuterated diazomethane was reacted with a mixture of surfactants (Figure 8). The order of elution is octyl sulfate, decyl sulfonate, dodecyl sulfate, dodecyl sulfonate, and a C_{12} phenyl sulfonate (mixture of isomers). Unlike the MPD 850, the sensitivity and selectivity of the oxygen channel response is roughly equivalent to that of the other elements. Sample stability and reproducibility are

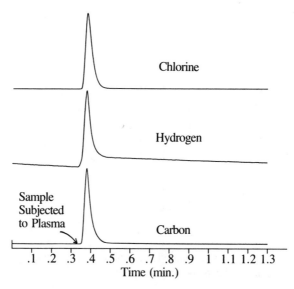

Figure 7. MPD response versus time from plasma pyrolysis of a PVC polymer illustrating chlorine (top), hydrogen (middle), and carbon response (bottom).

poor unless excess alcohol-free diazomethane is used to prevent alkylations and other side reactions.

The atomic emission detector has been described as an absolute detector, where element response is, for the most part, compound independent. To verify that, a packed column analysis of a mixture of CF_4, CH_4, SF_6, C_2F_4, C_3F_8, SO_2, and CS_2 was run (Figure 9). It is readily apparent from the carbon-selective chromatogram that, without a selective detector, the latter two compounds would be very difficult to detect due to significant column bleed. In addition, although CF_4 and CH_4 are not totally resolved, it is possible to quantitate for either using the fluorine or hydrogen channels of the AED. In order to demonstrate the compound independent response of the AED, calibration mixtures were prepared using several of these compounds (e.g. the source of the sulfur response for a given calibration point could be any of the sulfur-containing materials, like CS_2, whereas the next point could represent sulfur response from SO_2). The calibration curves for carbon, sulfur, and fluorine are linear and compound independent. Our most recent work indicates that there may be some variability in fluorine response if SF_6 is used for the source for fluorine; this effect appears to be scavenge gas-dependent. This system has been used to monitor a broad range of gas phase reactions, and where standards are not available for all of the components of a mixture.

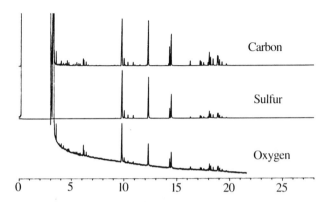

Figure 8. Chromatograms of methylated surfactants illustrating carbon (top), sulfur (middle), and oxygen (bottom) response.

Figure 10 illustrates the analysis for a sulfur-containing additive in a polymer matrix. A solvent extract of the sample yielded a complex carbon chromatogram indicating the presence of significant co-extractants as well as some interference from column bleed. The sulfur channel is free from interferences and presents a clear picture of the oligomeric composition of the additive. Area summation allowed for quantitation, without further attempts at sample cleanup (10).

A mixture of ten pesticides (vernolate (peak 1), atrazine (peak 2), fonofos (3), diazinon (4), metribuzin (5), alachlor (6), heptachlor (7), aldrin (8), Endosulfan I (9) and sulprofos (10)) was spiked into one liter of surface water at the 1 ppb level. The water was filtered through a 3M Empore™ membrane containing a C_8-bonded silica particulate enmeshed in fibrillated Teflon™. The pesticides were concentrated out of the water in a very clean form of solid phase extraction. Elution with ethyl acetate and concentration to a final volume of one ml resulted in a carbon trace complicated by coextractants. The sulfur, phosphorous, and chlorine chromatograms obtained from the AED (Figure 11) would allow for quantitation of the pesticides without further sample cleanup and facilitate qualitative identification of those pesticides containing specific combinations of heteroelements. The determination of phosphorous is enhanced using a hydrogen scavenge gas and necessitates an increase in plasma gas flow to reduce tailing.

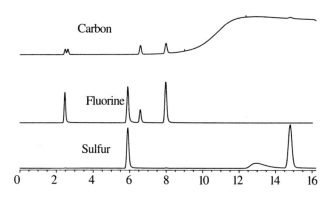

Figure 9. Chromatograms of gas standards on a packed column, carbon (top), fluorine (middle), and sulfur (bottom) response.

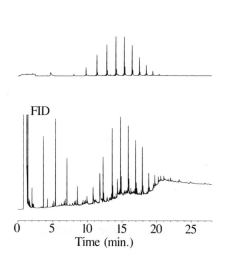

Figure 10. Chromatograms of a polymer extract, FID and sulfur response.

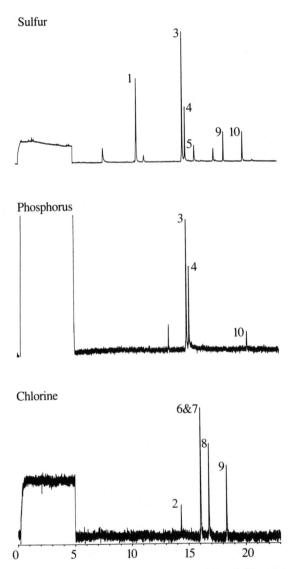

Figure 11. Chromatograms of pesticides recovered from surface water at the ppb level, sulfur (top), phosphorous (middle), and chlorine (bottom) response.

Acknowledgments. We would like to thank Dr. Louis C. Haddad, Jon Belisle, Elizabeth Sullivan, Richard Rossiter, and Vicki Bunnelle for their contributions to the analyses described above.

Literature Cited.

1. Uden, P. C. *Trends Anal. Chem.* **1987**, *6*, 238-246.
2. Uden, P. C.; Yoo, Y.; Wang, T.; Cheng, Z. *J.Chrom.* **1989**, *468*, 319-328.
3. Ebdon, L.; Hill, S.; Ward, R. W. *Analyst* **1986**, *111*, 1113-1138.
4. Hagen, D. F.; Belisle, J.; Marhevka, J. S. *Spectrochim. Acta* **1983**, *38B*, 377-385.
5. Hagen, D. F.; Belisle, J.; Johnson, J. D.; Venkateswarlu, P. *Anal. Biochem.* **1981**, *118*, 336-343.
6. Hagen, D. F.; Marhevka, J. S.; Haddad, L. C. *Spectrochim. Acta* **1985**, *40B*, 335-347.
7. Hagen, D. F.; Haddad, L. C.; Marhevka, J. S. *Spectrochim. Acta* **1987**, *42B*, 253-267.
8. Beroza, M.; Coad, R. A. *J. Gas Chromatog.* **1966**, *4*, 199-216.
9. Hagen, D. F.; Marhevka, J. S. *U.S. Patent 4,532,219*, **1985**.
10. Sullivan, E. A.; Rossiter, R. C. *H.P. Application Note 228-102*, **1989**.

RECEIVED May 6, 1991

Chapter 8

Analytical Problem Solving with Simultaneous Atomic Emission–Mass Spectrometric Detection for Gas Chromatography

D. B. Hooker and J. DeZwaan

Physical and Analytical Chemistry Research, 7255–209–0,
Upjohn Company, Kalamazoo, MI 49001

Atomic emission generated when an organic sample is introduced into a microwave induced plasma is useful analytically because of the efficiency with which many important elements can be excited, and the selectivity and linearity of the spectrally resolved emission signals. The elemental selectivity of the technique allows materials incorporating elements of interest to be readily located in complex media, while the linearity of the response enables quantitative analysis, since equal detector response will be obtained for each element independent of its parent chemical structure. This utility can be expanded if several elemental responses are monitored simultaneously, along with the mass spectral data for each material eluted from the sample. Often this additional information not only allows the materials of interest to be located and quantitated but also allows the molecular formulas of unknown materials to be determined, frequently without requiring instrument calibration. This capability makes the combined AED-MSD detector both a powerful means of molecular structure determination, and an excellent quantitative tool with high specificity.

The need to locate, identify or accurately quantify materials in a mixture is a situation frequently encountered by analytical chemists involved in chemical or pharmaceutical research. The matrices encountered in these research programs can vary from fairly simple, as in the determination of trace impurities in a bulk material, to very complex, as in the search for metabolites in biological media or extracts. Because of its high chromatographic efficiency, modest sample size requirements, and ease of operation, capillary column gas chromatography is a preferred method of analysis when it can be applied. Another advantage associated with gas chromatography is the capacity to employ an array of powerful detectors reliably and with a minimum of difficulty.

0097–6156/92/0479–0132$06.00/0
© 1992 American Chemical Society

These detectors, which include mass selective (MSD), infrared (IR) and atomic emission (AED), can provide the additional selectivity, structural information or equal detector response essential in resolving a variety of complex analytical problems.

A microwave induced plasma (MIP) sustained in a helium matrix, is especially applicable for chemical or pharmaceutical research because it can efficiently excite a variety of non-metals such as carbon, hydrogen, deuterium, nitrogen, oxygen, sulfur, phosphorus and the halogens. Since the MIP emission technique was first demonstrated by McCormack et.al.(*1*), it has been extensively evaluated as an AED for gas chromatography (*2-6*). Microwave induced plasmas can be generated at a variety of helium pressures and using various resonant cavities or wavelaunchers (*7-10*). Several of the common experimental arrangements have produced responses that are 1) linear over wide ranges of sample size, 2) elementally selective and 3) reasonably sensitive (*11,12*). Because emission from all elements introduced into the plasma occur at the same time, it is possible, with the proper optical arrangement, to monitor several elemental responses simultaneously. The elemental selectivity which is available with an MIP can be used either to locate a material of interest in a complex mixture or to spectrally resolve, by element, materials which have not been chromatographically resolved. Since the elemental responses produced are substantially independent of the chemical environment, this detector provides equal detector response for all compounds which contain a particular element, hence complete or partial empirical formulas can be determined for eluted materials. These may often be obtained with minimal instrument calibration.

Although the data generated by an MIP emission detector are similar in many ways to those produced by a mass selective detector, the information provided is in many instances complementary. For example, mass selective detectors give increased selectivity by monitoring a chromatogram for certain masses or for particular isotope patterns (as in the case of chlorine and bromine) but not particular elements. Therefore if the species of interest (or a mass fragment of that species) is known in advance or an element with a characteristic isotope pattern is known to be present, MSDs can often give excellent selectivity. AED detectors do not require any previous knowledge of the structures of the materials. To be useful as a selective detector, however, the species of interest must contain an element that is not common to components of the matrix. For quantitating materials of unknown structure, total ion current (TIC) responses are often misleading on an area percent basis, while an AED gives a response directly proportional to the elemental content of a material, independent of structure. With respect to structure determination, the MSD gives information on the molecular weight of the material and/or major fragment ions while an AED can give information on its elemental composition.

The use of a combined MSD-AED detector allows for the convenient collection and correlation of both data sets. The construction of a combined detector, using both a low pressure and an atmospheric pressure splitter, and its operating characteristics are described. The use of this combined detector in a variety of problems is also discussed for materials of many different elemental compositions. These problems include the exact determination of molecular formulas, the characterization and identification of low level impurities in bulk chemicals, the analysis of reaction mixtures and the location and characterization of metabolites in biological matrices.

Experimental Section

The combination detector used in generating the MIP-MS data is based entirely on commercial instrumentation. The capillary column gas chromatograph was a model 5890 (Hewlett-Packard, Avondale,PA) which was interfaced to a Hewlett-Packard series 5970B mass selective detector. A Hewlett-Packard model 59970A workstation was used to control these instruments as well as to accumulate and store the mass spectral data. All chromatograms reported were generated using a DB-5, 15 meter, 0.25 mm ID, fused silica column (J&W Scientific, Folsom,CA).

The MIP emission spectrometer used was a model MPD 850 (Applied Chromatography Systems, Luton, England). This instrument incorporates a low pressure (5-10 torr) 1/4 wavelength Evenson type (7) cavity and a 3/4 meter (dispersion = 1.39 nm/mm) monochromator. This is a non-scanning monochromator with several fixed exit slits which can be monitored. The emission wavelengths given in Table I are simultaneously monitored using two four-channel interface boxes associated with a MAXIMA 820 Chromatography Workstation (Dynamic Solutions, Ventura CA). The Evenson cavity system was used unmodified to generate the MIP, except for the flow system which delivers a helium-scavenge gas mixture to the plasma. This incorporates a flow controller, (Model 200-SSVB2, Porter Instrument Company, Inc., Hatfield, PA) used to deliver known amounts of scavenge gas to a mixing volume where it is mixed with chromatographic grade helium. The mixed gases are drawn through a restrictor which delivers 0.5 ml/min of gas, the excess being vented to the atmosphere. The helium-scavenge gas mixture (0.5 ml/min) is recombined with pure helium (25 ml/min) in a final mixing volume and is introduced into one of the channels of the transfer line (described below) inside the gas chromatograph oven. This arrangement, which allows the type and level of scavenge gas in the plasma-sustaining helium to be rapidly and reproducibly varied, extends the versatility and stability of the detector.

The transfer line which delivers both the sample and the helium feed (spiked with scavenge gas) to the plasma consists of three concentric tubes wrapped in heating tape. The inner tube (0.48 mm OD by 0.22 mm ID aluminum clad vitreous silica) contains the GC effluent after splitting. The middle tube (1/16 inch OD by 0.040 inch ID stainless steel) conducts the helium feed to the plasma. The outer 1/4 inch copper tube provides mechanical stability and is wrapped with heating tape. The temperature of the transfer line is monitored by an iron-constantan thermocouple and maintained at a temperature greater than the GC oven temperature (usually 275 degrees Centigrade).

The chromatographic effluent is split inside the chromatograph oven. This sample split can be accomplished at reduced pressure as described in detail in reference 13, or at atmospheric pressure as described here. The atmospheric pressure system has the advantages of 1) producing chromatograms with retention times identical to those obtained using an FID or other atmospheric pressure detector and 2) being less susceptible to air leaks which complicate the determinations of the nitrogen and oxygen contents of materials eluted.

The atmospheric splitter used is shown in Figure 1. The GC effluent is introduced at the bottom of the tube (Figure 1b) at a rate of approximately 1 ml/min through the capillary column (F). Because the entire assembly is at atmospheric pressure through vent (D), mass flow to the mass selective detector must be limited in some way to maintain the instrument pressure within its optimal operating range. In this splitter the impedance to mass transfer is provided by the viscous flow through a narrow diameter capillary tube (E). The optimum length of a capillary restrictor of given diameter depends on the temperature range to be used for the chromatograph oven, because of the increase in gas viscosity that occurs with increasing temperature. Effluent that is not taken into the mass selective detector is taken up by tube (G) and is introduced into the MIP. Because the inputs to the detectors are removing more material from the tube than is being introduced by the chromatograph, helium make-up must be provided through tube (C) to prevent air from being drawn into the tube. The flow of this make-up gas is carefully regulated to provide for only a slight flow of helium to vent though tube (D). This is a critical adjustment because too slow a flow will allow air into the system while too fast a flow will result in convection and mixing inside the apparatus. The adjustment of this flow has been best made by gradually increasing make-up flow until significant nitrogen signals are no longer observed on either detector.

Instrument Characteristics

Obtaining both atomic emission data from several elements and mass spectral data from a chromatographic sample can easily result in a large accumulation of data. In simple experimental situations it may be possible to correlate data obtained from separate MIP and MS chromatograms. In complex mixtures, containing several minor components of interest, small shifts in retention times can be caused by slightly different vacuums, temperature programs or sample loadings, even if the same column is used on two different instruments. To be useful in complex mixtures it is necessary that responses obtained from the MIP emission be directly correlated with responses from the mass analyzer. This correlation can be readily achieved using the combined detector even in complex samples as illustrated by the carbon emission and total ion current responses in Figure 2, which were obtained simultaneously using the atmospheric pressure splitter. Also shown in Figure 2 is the FID response obtained from the same sample analyzed under identical conditions using the same column on a different instrument. The major differences between the chromatograms in Figure 2 arise from differences in the relative peak responses between detectors and in presentation differences inherent in the data systems. No significant differences in retention times are observed, although a slight increase in the peak widths is apparent in the MIP-MSD data.

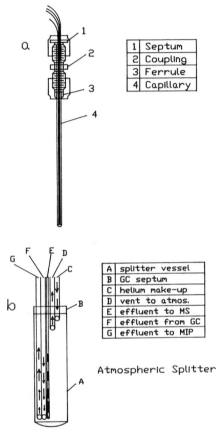

1	Septum
2	Coupling
3	Ferrule
4	Capillary

A	splitter vessel
B	GC septum
C	helium make-up
D	vent to atmos.
E	effluent to MS
F	effluent from GC
G	effluent to MIP

Atmospheric Splitter

Figure 1. a) Overall view of the atmospheric splitter with a GC septum cut to size (1), a 1/16" Parker Coupling (2), a 1/16" teflon ferrule (3), and a size 99 Kimble (Kimax-51 #34505)closed end capillary (4).   **b)** Close-up view showing the gas flows. C, D and G are aluminum clad deactivated silica tubing (0.22 mm ID from Scientific Glass Engineering #061220) while E is 0.1 mm ID polyimide clad, deactivated silica (SGE#062469).

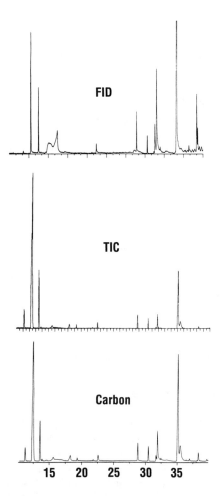

Figure 2. The TIC, carbon emission and FID responses showing trace impurities (peak at 12.5 minutes was spiked at 0.2%) in a bulk material. The TIC and carbon responses were obtained from the same sample injection but the FID response was obtained from the same column on a different instrument. Major peak elutes at 40 minutes.

Presented in Table I are detection limits of this combined detector for various elements, the values reported reflecting the weights loaded on the column before the sample was split. These values are the amounts of material required to produce a signal which is twice the peak to peak noise, and were determined by injecting successively more dilute samples until the response obtained for the element became of the same order of magnitude as the noise in that channel. The values are comparable to those determined for this type of microwave plasma by previous investigators.

Table I. Emission Wavelengths, Sensitivities and Selectivities Obtained Using Combined AES-MSD Detector

Element	Wavelength (nm)	Sensititvity (ng/sec)	Molar Selectivity (Mole/Mole Carbon)
Carbon *	247.9	0.082	-----
Hydrogen*	486.1	0.043	-----
Fluorine*	685.6	0.40	350
Chlorine*	479.5	0.36	200
Sulfur*	545.4	0.55	170
Deuterium*	656.1	0.02	80
Phosphorus**	253.6		
Oxygen***	777.2		

*   Determined using 0.2% oxygen as scavenge gas
**   Determined using 0.2% hydrogen as scavenge gas
*** Determined using 0.2% of a 95:5 nitrogen:hydrogen gas mixture as scavenge gas

The utility of the MIP emission data arises primarily from two unique capabilities. The first is the elemental selectivity which can be obtained and the second is the equal detector response provided for each individual element which can be monitored. The elemental emission responses obtained using an instrument with fixed slits and a non-scanning monochromator, include both the desired emission (if there is any) and the ever present and unwanted background which arises from low levels of broad band non-atomic emission which occurs when samples are introduced into the plasma. The magnitude of these background emissions are usually related to the magnitude of the associated carbon signal and it is these "ghost" signals which are the factor limiting elemental selectivity in this optical arrangement. The selectivity ultimately obtained is a function of many variables, such as: nature of scavenge gas, level of scavenge gas, microwave power level, type of microwave cavity and helium pressure. Elemental selectivities typically obtained using this instrument with 100 watts forward power and 0.2% oxygen scavenge gas are given in Table I. These numbers should be interpreted as the number of moles of carbon required to produce a signal, equal in magnitude to a single mole of the specified element at the emission wavelength of that element.

Presented in Table II are the structures and sources of a series of compounds,

Table II. Materials Used Along with Their Source, Molecular Formulas, and Some Experimental Empirical Formulas Are Given in Parenthesis.

Compound	Number	Formula	Source
1-chloro-2-fluorobenzene	1	C_6H_4ClF	Aldrich
3-chloroanaline	2	C_6H_6NCl	Aldrich
Benzothiazole	3	C_7H_5NS	Eastman
2-chloro-benzothiazole	4	C_7H_4ClNS	Eastman
2-fluorobiphenyl	5	$C_{12}H_9F$	Aldrich
Benzophenone	6	$C_{13}H_{10}O$	Aldrich
Diphenylether	7	$C_{12}H_{10}O$	Aldrich

| | 8 | $C_{16}H_{21}N$ ($C_{13.6}H_{18.7}N$) | UPJOHN |

| | 9 | $C_{19}H_{18}N_5Cl$ ($C_{18.4}H_{18.9}N_{5.5}Cl$) | UPJOHN |

| | 10 | $C_{16}H_{22}N_2Cl_2O$ ($C_{15.7}H_{22.6}N_{2.2}Cl_{2.1}O$) | UPJOHN |

| | 11 | $C_{18}H_{14}NO_2SF_3$ ($C_{19.2}H_{14.8}N_{1.3}O_{2.9}F_{3.2}S$) | UPJOHN |

SOURCE: Reprinted with permission from Reference 15. Copyright 1989

whose determined elemental response ratios and their standard deviations are given in Table III. These response ratios (X/C) were determined from the individual integrated elemental responses by using the following relation -

$$\frac{X}{C} = \frac{R_x}{R_c} \frac{N_c}{N_x}$$

In this expression R_x and R_c represent the areas observed for the element, X, and for carbon respectively, and N_x and N_c are the number of X atoms and carbon atoms in the empirical formulas of the compounds. No corrections of any kind (14) were made to the raw data in generating the values in Table III and the precision of the numbers reported is typical of data taken on the same day. Because the values

---

Table III. Atomic Emission Responses Observed Relative to the Corresponding Carbon Responses for the Compounds as Numbered in Table II.

	Hydrogen	Nitrogen	Oxygen	Chlorine	Sulfur	Fluorine
1	0.698(0.031)	----	----	4.26(0.11)	----	6.11(0.18)
2	0.640(0.011)	2.40(0.09)	----	4.47(0.15)	----	----
3	0.602(0.015)	2.52(0.17)	----	----	4.55(0.14)	----
4	0.565(0.017)	2.23(0.08)	----	4.44(0.10)	4.72(0.11)	----
5	0.674(0.012)	----	----	----	4.83(0.10)	----
6	----	----	1.35(0.04)	----	----	----
7	----	----	1.36(0.05)	----	----	----
8	0.665(0.016)	2.80(0.43)	----	----	----	----
9	0.687(0.021)	2.69(0.32)	----	4.66(0.24)	----	----
10	0.662(0.008)	2.66(0.24)	1.39(0.07)	4.65(0.01)	----	----
11	0.628(0.016)	2.81(0.16)	1.88(0.07)	----	4.35(0.12)	6.04(0.23)
ave.	.646(0.042)	2.59(0.22)	1.50(0.26)	4.57(0.11)	4.54(0.19)	5.99(0.13)

SOURCE: Reprinted with permission from reference 15. Copyright 1989.

NOTE: Numbers represent an average of five responses with the standard deviations given in parenthesis. All determinations were run using 0.2% oxygen scavenge gas except for oxygen which was run in 0.2% of a 95:5 nitrogen: hydrogen mixture.

---

reported in Table III depend directly on various instrument settings, they do not represent the relative sensitivities of the elements (see Table I), but they do demonstrate the relatively good compound-to-compound reproducibility necessary to determine empirical formulas. The experimental empirical formulas reported in Table II were determined by using the compound average of relative response ratios together with the individual element response integrals of each compound. If the molecular weights of compounds 8, 9, and 10 are known, the exact molecular formulas of these materials can be determined from these experimental empirical formulas as described in reference *15*. For compound 11, knowledge of the molecular weight can limit the possible molecular formulas to three. This compares to the 23 possibilities for a molecular formula which are within 5 millimass units of the exact mass calculated for this compound.

Applications

The selectivity, linearity, equal detector response and molecular weight capabilities of this detector, may be used in various combinations to approach a variety of common problems ranging from locating and identifying metabolites in biological matrices to rapid identification of impurities in bulk materials or the analysis of reaction mixtures.

An example of the utility of equal detector response is provided by the example in Figure 2, where the responses from three detectors for the same sample may be compared directly. This chromatogram shows the early eluting impurities present in a bulk drug lot. The major component itself begins eluting at 40 minutes and is not shown, while the peak eluting at 12.5 minutes is 2-chlorobenzothiazole which was added as an internal standard at a level of 0.2%. Comparing the responses of the various detectors relative to the internal standard peak clearly shows that different values of impurity levels would be obtained from the different detector responses. Based on the linearity and the compound-to-compound consistency of response that characterize the MIP carbon emission, this detector should provide the best direct comparison of impurity levels for materials of unknown structure.

Identification of Impurities.   As illustrated above, purity determinations are one of the most common applications of gas chromatography. One of the simplest and most frequent applications of GC-MS is in identifying the impurities detected. If the materials under investigation are common, it is very likely that minor components detected can be easily identified from their mass spectrum by using the extensive libraries available. If, however, newly synthesized materials are being evaluated, it is unlikely that relevant mass spectra are in standard libraries, and structure elucidation becomes more difficult. In these difficult cases the complementary data obtained from MIP emission may be extremely useful in structure determination, as well as, in the accurate quantitation of the relative impurity levels.

Shown in Figure 3 are the responses obtained in the carbon, chlorine, and fluorine emission channels, as well as the total ion current (TIC) chromatogram for a sample of a lot of compound 11 from Table II which had been synthesized by an outside laboratory to incorporate carbon-13 labels. From the carbon responses it appears that the material (peak B) contained two impurities (peaks A and C) at about

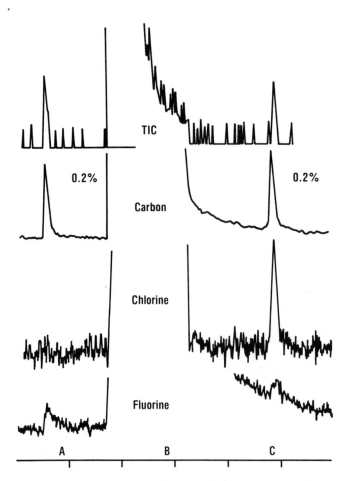

Figure 3. The TIC, carbon, chlorine and fluorine responses from a bulk sample of compound 11 (Table II), showing the major component (Peak B) and two small process impurities. Reprinted with permission from reference *13*. Copyright 1989 Pergamon Press, plc.

the 0.2% level. Because these particular impurities had never been observed in other lots of this compound, it was essential to identify them. Given in Table IV is additional information gathered from this same data set, including the molecular weights observed and the F/C, S/C and Cl/C response ratios observed for each individual peak. Direct comparison of these elemental ratios is possible, even though the magnitude of the peak B responses were 500 times larger than those for peaks A and C, because of the excellent linearity of the MIP emission detector. The data in Table IV indicate that compound A has nearly the same level of fluorine as peak B and also, like peak B, it contains no chlorine. Unlike peak B, however, it contains no sulfur and has a molecular weight which is 16 mass units lower than B. These facts and the very similar chromatographic behavior make the oxazole structure given in Table IV a very likely possibility for peak A. The emission data for compound C indicate that sulfur is present at about the same level as in B but that the relative fluorine level is diminished and that chlorine is definitely present. Consideration of this elemental composition information along with the increase of 16 in the observed molecular weight of compound C, indicates that it is the difluorochloromethyl material indicated.

Table IV. Elemental Response Ratios, Molecular Weights and Proposed Structures for Components A ,B and C Defined in Figure 3.

Response	Peak A	Peak B	Peak C
F/C Ratio	0.77 (0.12)	0.83 (0.05)	0.43 (0.04)
S/C Ratio	0.006 (0.005)	0.11 (0.01)	0.13 (0.00)
Cl/C Ratio	0.06 (0.004)	0.05 (0.001)	0.69 (0.06)
GMW	355	371	387
Structure			

SOURCE: Reprinted with permission from reference 13. Copyright 1989.

In the case above, the combination of molecular weight and limited elemental composition data allowed a problem to be completely resolved in a couple of hours. This was accomplished without any external calibration and required only that the sample size used produced a measurable response for all elements of each material. The fact that the sensitivity of the emission detector was similar for all of the elements of interest (S, F and Cl) was helpful. In situations where data are required for elements giving less sensitivity (N or O), additional sample preparation may be needed to increase the concentration of the impurity relative to the major component. This concentration can often be done by exposing a large amount of the sample to a relatively small amount of a solvent as described in reference *16*. The solution soon becomes saturated in the main component, but will keep dissolving impurities, thus concentrating the impurities in the solution relative to the bulk. Enrichments of impurity levels of an order of magnitude or more may often be realized by employing this procedure.

Shown in Figure 4 are the carbon, hydrogen, chlorine and nitrogen responses obtained from a sample in which the level of a process impurity present in a lot of bulk chemical (compound 10 in Table II) had been enriched. This enrichment procedure allows the nitrogen and oxygen contents of the small component to be determined without grossly overloading the column. Using the responses of the parent material to calibrate the responses of the impurity, the elemental ratios determined for the impurity are given in Table V. Since the molecular weight of the impurity is even, as indicated by the mass spectrum in Figure 5, and nitrogen is present in the molecule, there must be two nitrogens in the material. Using the H/C, Cl/C and N/C ratios in Table V, which were determined simultaneously with oxygen scavenge, and the molecular weight of 272, the molecular formula of the impurity must be either $C_{12}H_{14}Cl_2N_2O$ or $C_{13}H_{18}Cl_2N_2$. More information is required, however, to establish the molecular formula unambiguously. This information was obtained in the form of the O/C ratios given in Table V, but required an additional sample injection for determining the carbon and oxygen responses using a 5% hydrogen-95% nitrogen scavenge gas. The fact that the material contains no oxygen, confirms the molecular formula of $C_{13}H_{18}Cl_2N_2$ and the structure as the secondary amine (not the propanamide).

Table V. Experimental Area Response Ratios for the Parent and Impurity Peaks of Figure 4 and the Known Elemental Ratios of the Parent and the Calculated Elemental Ratios of the Impurity.

Experimental Response Ratios				Elemental Ratios			
		Parent Material ($C_{16}H_{22}Cl_2N_2O$)					
H/C*	Cl/C*	N/C*	O/C**	C/H	C/Cl	C/N	C/O
0.517	0.129	0.135	0.089	0.727	8.00	8.00	16.0
(0.002)	(0.005)	(0.001)	(0.04)				
		Process Impurity					
0.519	0.183	0.171	0.000	0.730	6.25	5.55	-----

* Determined with O_2 scavenge gas. Average of five determinations with the standard deviations given in parenthesis.

**Determined with 95:5 Nitrogen:Hydrogen scavenge gas.

Analysis of Reaction Mixtures. The analysis of reaction mixtures, especially in areas where the chemistry involved is not well understood, is another useful application of gas chromatography with sophisticated detectors. An illustration of this

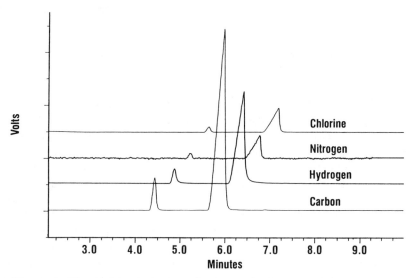

Figure 4. The carbon, hydrogen, nitrogen and chlorine emission responses of a sample of compound 10 (Table II) in which the level of the impurity has been enriched.

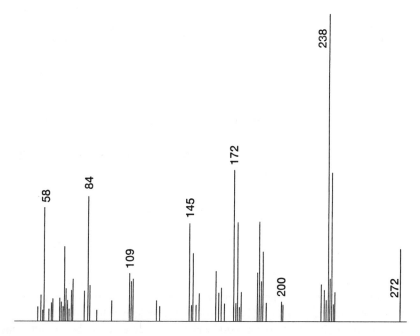

Figure 5. Mass spectrum of the impurity peak obtained simultaneously with the emission responses in Figure 4.

involving, phosphorus chemistry, is provided by the reaction products obtained from the following reaction.

$$\langle \bigcirc \rangle\text{—C}\!=\!\text{CH}\overset{\overset{\displaystyle O}{\|}}{P}(OMe)_2 \quad + \quad (MeO)_2\overset{\overset{\displaystyle O}{\|}}{P}H$$

The TIC, carbon, oxygen and phosphorus responses obtained from the gas chromatogram obtained from this reaction mixture are given in Figure 6. It is clear that the mixture has at least 5 components, each of which contains carbon, oxygen and phosphorus. The mass spectrum of peak A provides a molecular ion of 212 daltons and indicates that A is the starting material with molecular formula $C_{10}H_{13}PO_3$. The ratios of the area responses obtained for the individual elemental responses for the various peaks are given in Table VI. Because of the tailing observed in the phosphorus response when the 95% nitrogen - 5% hydrogen scavenge gas was used, the elemental response ratios for phosphorus were determined using hydrogen as scavenge gas. Using the values of area response ratios determined for peak A and the known composition of this material to calibrate the observed responses, the elemental ratios of the peaks of unknown structure may be calculated. These element ratios and the empirical formulas consistent with them are also given in Table VI.

As discussed in reference *16*, knowledge of the molecular weights of the materials allows the experimentally determined empirical formulas to be refined to give the exact molecular formulas in most cases, because the molecular formulas must be consistent with the molecular weights and also must contain only integer levels of the individual elements. These constraints lead to the molecular formulas given in Table VI for peaks B, D and E for which molecular ions were obtained. The molecular ion for peak C was not found since the highest mass peak observed was 138, which is much too low to be consistent with the empirical formula found for peak C and therefore it is not possible to determine the exact molecular formula for C.

The molecular formula found for peak D, $C_{12}H_{20}P_2O_6$, is consistent with the value expected for the desired reaction product:

$$\langle \bigcirc \rangle\text{—CH—CH}_2\text{—}\overset{\overset{\displaystyle O}{\|}}{P}\text{—(OMe)}_2 \\ \qquad \quad \underset{\displaystyle O\!=\!P\text{—(OMe)}_2}{|}$$

The molecular formula found for peak E, $C_{18}H_{15}PO$, was unexpected since it appeared to contain only a single oxygen and because a reasonable structure arising from the initial reagents could not be proposed. The molecular formula, however, is consistent with triphenyl phosphine oxide which was later shown to be present in the starting material.

Table VI.   Experimental Elemental Ratios, Empirical Formulas, Molecular Weights and Molecular Formulas Determined from Reaction Mixture of Phosphorus Compounds as Described in Figure 6.

		Element Ratios			Formulas	
Peak	GMW	C/P	C/O	C/H	Empirical	Molecular
A	212	10.0	3.33	0.77	$C_{10}H_{13}PO_3$*	$C_{10}H_{13}PO_3$
B	242	14.3	6.67	0.65	$C_{14.3}H_{22}PO_{2.1}$	$C_{13}H_{23}PO_2$
C	(138)	13.5	3.4	0.65	$C_{13.5}H_{20.8}PO_{3.97}$	Unknown
D	322	5.7	2.0	0.57	$C_{11.4}H_{20.4}P_2O_{5.7}$	$C_{12}H_{20}P_2O_6$
E	278	17.5	17.0	1.28	$C_{17.5}H_{13.7}PO$	$C_{18}H_{15}PO$

* Known values used to calibrate the responses.

The significant impact of scavenge gas composition on the responses observed for the elements involved in this example demonstrates the importance of being able to shift rapidly and reliably from one scavenge gas to another. This practice does, however, require multiple sample injections per sample.

Selectivity in Complex Matrices.   In situations where the compound or family of compounds of interest is known to contain an element which is not commonly found in the matrix being studied, the MIP emission detector can be very useful in locating them. In biological samples, for example, compounds containing chlorine, fluorine, sulfur, bromine or phosphorus are effectively labelled and can often be picked out of complex matrices. An example of locating a fluorine containing molecule (compound 11 from Table II) in a very complex matrix (extracted blood plasma) is presented in Figure 7. The sample was obtained from a rat which had been dosed intravenously with the compound. This example demonstrates that, with an appropriate element present, materials of interest can be readily located in this complex mixture at typical blood levels (μg/ml), in the presence of much larger ghost peaks, by comparing the relative responses in the carbon and fluorine channels. The mass spectrum associated with the fluorine containing peak is shown in Figure 8. Because the two data sets were obtained simultaneously, no problem was encountered in correlating the emission and mass spectral data sets even for a minor component in a complex matrix. The chemical structures of materials may often be readily determined from their mass spectra. This is particularly true in cases like biological metabolites, where a series of labelled compounds may be generated from a parent compound of known structure.

Since the presence of an appropriate element in the parent molecule of interest can not be assured, the use of MIP emission as a selective detector may not be readily accomplished in all cases. Synthetically modifying materials to include elements such as sulfur or halogens would almost certainly dramatically alter their biological activity and therefore this approach is not useful for labelling materials. Methods of synthetically incorporating elemental labels into materials prior to

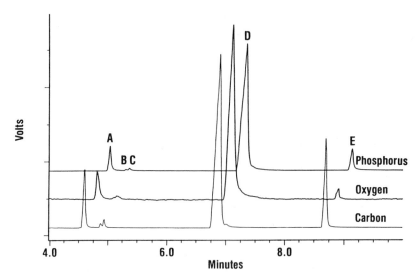

Figure 6. The carbon and oxygen responses observed with a 95% nitrogen-5% hydrogen scavenge gas for the reaction mixture of phosphorus containing materials described. Also shown is the phosphorus response obtained using a hydrogen scavenge gas.

chromatographic analysis have been demonstrated (*17,18*) and may be practical in some cases. Procedures of this type, however, may either fail to derivatize certain structures or label components of the matrix in complex samples. A more reliable method of labelling is by incorporating stable isotopes, such as deuterium, directly into the parent material. The data in Table I indicate that good sensitivity can be obtained for deuterium from the AES detector. Presented in Figure 9 are the carbon and deuterium emission responses obtained from a sample of extracted blood serum. This serum was obtained from a rat which had been dosed intravenously with a selectively deuterated benzodiazapine. Comparison of the relative intensities of the carbon and deuterium responses easily allow the peaks associated with the parent material and its major metabolite (indicated in Figure 9 with stars) to be located even in the presence of large ghosting peaks associated with major components of the matrix.

Although the combined AED-MSD detector can in many cases be very useful in the location and characterization of metabolites, it is not as generally powerful and useful as the common procedure of using radioisotopes with HPLC analysis. This is because of the necessity of getting the samples through a gas chromatograph. The process of metabolism generally acts to increase the polarity of the material and thus decreases its compatibility with GC. Therefore, even in cases where the parent molecule can readily be analyzed by GC, it is very likely that some of its metabolites are either not amenable to GC analysis or decompose in the injector. However, in early phases of research or development, before synthesis of radiolabelled materials may be practical, a great deal of useful metabolism information can often be obtained.

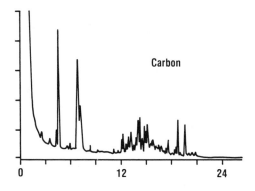

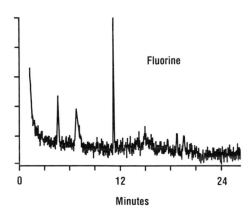

Figure 7. Carbon and fluorine MIP emission responses obtained for a chromatogram of an extracted sample of blood plasma obtained from a rat dosed with compound 11 from Table II. Reprinted with permission from reference *15*. Copyright 1989 by the American Chemical Society.

Conclusion

Atomic emission and mass spectral data provide complementary types of information. An AES gas chromatographic detector provides (1) equal detector response for all components, as well as excellent linearity, which allows accurate relative quantitation of mixtures, (2) the possibility of complete characterization of the elemental composition of all eluted materials and (3) element selectivity, while an MSD can provide molecular weight and fragmentation information and mass selectivity. The combination of both of these data sets provides the potential for application in a wide range of analytical problems.

A versatile and reliable gas chromatographic detector, capable of simultaneously generating atomic emission and mass spectral information for each component eluted from the column, was produced using commercially available equipment. This was accomplished using a low pressure MIP which, along with the

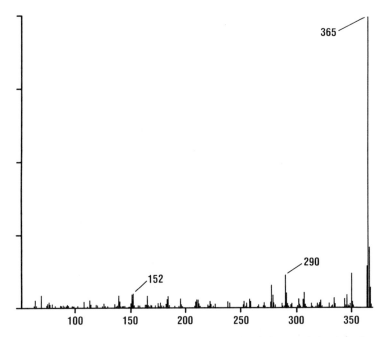

Figure 8. Mass spectrum of the fluorine containing peak in the plasma sample, obtained simultaneously with the emission responses in Figure 7. Reprinted with permission from reference *15*. Copyright 1989 by the American Chemical Society.

MSD inlet, served as a driving force to induce mass flow from the exit of the gas chromatographic column to the respective detector. This Evenson type resonant cavity also allows a wide range of elements to be effectively observed especially if the composition and level of the scavenge gas used can be readily controlled. The reliability and the convenient routine use of this combination detector require the design of a rugged sample splitter and transfer line. The atmospheric sample splitter described is not only reliable and easy to construct, but it also allows direct retention time comparison of the chromatograms generated using this detector with other atmospheric pressure detectors such as the common flame ionization detectors.

The analysis of the combined atomic emission - mass spectral data sets generated demonstrate the potential of this combined detector for elemental selectivity in complex matrices and for molecular formula determinations. These features can often be very useful in resolving complex problems faced in research situations where matrices are complex, samples are often limited and time is important. Although external calibration of each of the emission responses can be difficult and time consuming, this procedure is almost never required. This is because in situations where impurities are being identified or reaction mixtures are being analyzed, there is usually a peak present whose structure is known and can be used to calibrate the responses obtained for other materials sufficiently well to establish either their structure or their molecular formulas.

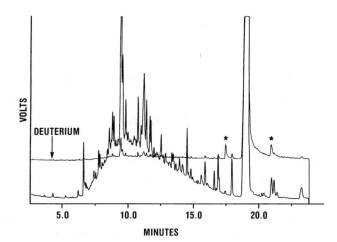

Figure 9. The carbon and deuterium emission responses for a chromatogram of extracted blood plasma taken from a rat dosed intravenously with a benzodiazepine enriched with deuterium. Comparing the relative magnitudes of the responses permits low levels of deuterated peaks to be located in the presence of large ghost responses. Reprinted with permission from reference *13*. Copyright 1989 Pergamon Press, plc.

Literature Cited

1. McCormack, A. J.; Tong, S. G.; Cooke, W. D. *Anal. Chem.* **1965**, 37, 1470.
2. McClean, W. R.; Stanton, D. L.; Penketh, G. E. *Analyst* **1973**, 98, 432.
3. Estes, S. A.; Uden, P. C.; Barnes, R. M. *Anal. Chem.***1981**, 53, 1829.
4. Brener, K. S. *J. of Chromatogr.* **1978**, 167, 365.
5. Bradley, C.; Carnahan, J. W. *Anal. Chem.* **1988**, 60, 858.
6. KehWei, Z.; Qing-Yu O.; Guo-Chuen, W.; Wei-Lu, Y. *Spectrochim. Acta* **1985**,40B, 349.
7. Fehsenfeld, F. C.; Evenson, K. M.; Broida, H. P. *Rev. Sci. Instrum.* **1965**, 36, 294.
8. Slatkavitz, K. J.; Uden, P. C.; Barnes, R. M. *J. Chromatogr.* **1986**, 355, 117.
9. Hubert, J.; Moisan, M.; Ricard, A. *Spectrochim. Acta* **1979**, 33B, 1.
10. Evans, J. C.; Olsen, K. B.; Sklarew, D. S. *Anal. Chim. Acta* **1987**, 194, 247.
11. Mohamad, A. H.; Caruso, J. A. *Advances in Chromatography* **1987**, 26, 191.
12. VanDalen, J. P.; De Lezenne Coulander, P.A.; DeGalan, L. *Anal. Chim. Acta.* **1977**, 94, 1.
13. Hooker, D. B.; DeZwaan, J. *J. Pharm. Biomed. Anal.* **1989**, 7, 1591.
14. Hass, D. L.; Caruso, J. A. *Anal. Chem.* **1985**, 57, 846.
15. Hooker, D. B.; DeZwaan, J. *Anal. Chem.* **1989**, 61, 2207.
16. Smith, G. B.; Downing, G. V. *Anal. Chem.* **1979**, 2290.
17. Hagen, D. F.; Marhevka, J. S.; Haddad, L. C. *Spectrochim. Acta* **1985**, 40B, 355.
18. Hagen, D. F.; Haddad, L. C.; Marhevka, J. S. *Spectrochim. Acta* **1987**, 42B, 253.

RECEIVED April 29, 1991

Chapter 9

An Element-Specific Detector for Gas Chromatography Based on a Novel Capacitively Coupled Plasma

B. Platzer[1], R. Gross[1], E. Leitner[1], A. Schalk[2], H. Sinabell[2], H. Zach[2], and G. Knapp[1]

[1]Department of Analytical Chemistry, Micro- and Radiochemistry, Graz University of Technology, Technikerstrasse 4, A–8010 Graz, Austria
[2]Anton Paar Company, A–8054 Graz, Austria

The Stabilized Capacitive Plasma (SCP) is a novel atomic emission source with stable and defined plasma geometry, good short and long term stability and robust plasma discharge. Radio frequency at 27.12 MHz is coupled into a water-cooled fused silica discharge tube by two annular electrodes. The RF-power dissipated in the plasma is approximately 140 W. Helium is chosen as plasma gas because of its high excitation efficiency for the halogens, but other gases, Ne, Ar, O_2, N_2 and CO_2 will form discharges in the SCP at atmospheric pressure. The instrument called ESD is an element specific detector for gas chromatography consisting of the SCP, an optical fiber for end-on observation of the atomic emission and conduction to an interference filter spectrometer for the simultaneous measurement of four different elements with background correction. The detection limits are in the low picogram-per-second range.

Today, there exists a demand for analytical methods and procedures with high selectivity and absolute power of detection down to the pg-range for numerous compounds in complex matrices. The combination of spectrometric and chromatographic techniques, so called "hyphenated methods" have become important and GC-MS and LC-MS are among the most selective available methods. A different approach for selective detection is coupling plasma atomic emission to gas chromatography (*1*).

Molecules eluting from the GC column are decomposed into their constituent atoms in a hot electrical discharge. These atoms are excited to emit element specific radiation. The wavelength of the emitted light is charac-

0097–6156/92/0479–0152$06.00/0

teristic of a certain element and the intensity of light is proportional to the number of atoms present in the discharge.

Notable advantages of plasma emission spectrometry coupled to capillary GC are listed below:
- all elements appearing in a GC peak can be measured selectively down to the picogram region. Therefore overlapping substances may be detected free of interferences, if the compounds contain different heteroelements.
- calibration is easier compared to other detector systems because only a low number of standards is needed for the calibration of many compounds of differing composition (2).
- with a suitable setup of the plasma emission detector, all elements of a separated compound can be measured simultaneously. In this way the molecular formula can be estimated (3).
- Deuterium has a distinct emission line. Therefore deuterated compounds can be distinguished from non-deuterated compounds (4).

These possibilities and prospects will greatly increase the importance of plasma emission detectors in chromatography in the near future. Many technical problems, that prevented widespread use of the method since its first introduction (5), have been overcome. For reliable quantitative measurements at the lowest signal levels, all of the components of such an instrument have to be carefully selected and matched to each other. Therefore the development of a plasma emission detector has to cover the following areas:
- Plasma sources (with an emphasis on a high temperature interface to the GC oven)
- Spectrometric systems (comprising light gathering optics, wavelength region of optical emission, spectrometers and optical sensors and methods for spectral background correction)

Plasma Sources

Modifications of the TM_{010}-MIP cavity as introduced by Beenakker (6) are the most frequently employed plasma excitation sources for nonmetals. Apart from improvements in energy coupling to the cavity and tuning capabilities many investigations focused on plasma torch geometry (7-9). It should be noted that the residence time of analytes in the detection volume can be increased only to an extent that it does not significantly contribute to peak broadening, an aspect not always taken into consideration (10).

During the last 6 years a new spectrochemical excitation source, the Stabilized Capacitive Plasma (SCP), has been developed in our laboratory. Before going into details of its construction and properties later in this paper, we would like to comment on related discharge configurations.

In the area of capacitively coupled microwave plasmas (CMP) research has concentrated above all on the problem of sample introduction into the plasma. A more recent arrangement uses a tantalum tube as electrode (*11*). A common characteristic of plasmas of this type is a metallic electrode in direct contact with the discharge, and the associated problems of electrode erosion and plasma contamination.

A helium-"afterglow" device as detector in capillary gas chromatography was introduced by Rice et al. (*12*). In the lower region of the discharge tube, gas discharge is induced with an annular electrode by means of high-voltage (several kV) applied at several hundred kHz. The excited species are swept downstream by the plasma gas. Metastable states have a lifetime which allows them to provide energy for the atomization and excitation of the sample molecules even a few centimetres above the visible discharge. The placement of an grounded counter electrode above the discharge, which was not used in earlier "true" afterglows by the same authors (*13*), does significantly change the apperance of the discharge (*12*). Obviously the afterglow is superseded by a capacitive discharge. Its low gas temperature is demonstrated by the fact that the GC capillary column protruding through the ring electrode is not destroyed.

Because of high purchasing and operating costs, the ICP is of limited significance in gas chromatography detection. A further disadvantage is that an argon plasma does not provide the excitation energy especially required by the halogens. The development of a 27 MHz He-ICP at atmospheric pressure by Chan et al. (*14*) showed, however, that in this source excitation of halogens and other nonmetals is very efficient.

Spectral Range

In the past, atomic emission spectrometry played a minor role in the determination of nonmetals, especially halogens. The reason was the unsuitability of most of the available spectrometers as regards to accessibility of appropriate spectral ranges (Vacuum-UV and especially near IR).

Houk et al. (*15*) showed that it is relatively easy to access the deep vacuum-UV range with an appropriate procedure. They immersed a sampler cone normally used in ICP-MS coupling directly into an ICP discharge. The sampler orifice was used as entrance slit for a purged VUV-spectrometer. In this way they were able to detect atomic emission below 100 nm. On the other end of the spectrum, the red and near infrared from 600 to 1200 nm, major problems arise from poor detector sensitivity (*16*). The portion from 900 to 1100 nm is especially inacessible to photomultipliers.

Freeman and Hieftje (*17*) combined a thermoelectrically cooled IR sensitive photomultiplier and a high throughput monochromator to compile useful emission spectra of nonmetals from an atmospheric pressure Helium MIP in a Beenakker cavity in the range from 700 to 1050 nm. They compared signal-to-background ratios for several elements in the NIR favorably with those in the UV/Vis range. Interferometric compilations of NIR atomic emission intensities of nonmetals were recorded by Pivonka et al. (*18*) for a high power Beenakker MIP and by Hubert et al. (*19*) for a Surfatron MIP. Line assignments show that most of the transmissions are nonresonant and therefore unlikely to show self absorption.

Commercial GC-Plasma AES Detector Systems.

The first plasma emission detector for GC was introduced to the market in the early 1970's by Applied Chromatography Systems Ltd. It was based on a low pressure MIP coupled to a Paschen Runge polychromator system. At that time the chromatographic community was not ready to accept such a complex and costly system as a GC-detector and it took more than 15 years before another plasma emission detector became available. This was developed by scientists at Hewlett Packard (*20*) and consists of a modified Beenakker cavity mechanically fixed to a concave grating spectrometer. The GC effluent is brought to the discharge by a heated flexible interface. Radiation from the water cooled discharge is focused into the spectrometer and detected by a photodiode-array (PDA). Because of the resolution-spectral range tradeoff only a small portion of the spectrum can be viewed at a given time. Only elements whose lines are in the same spectral window, can be detected simultaneously. The output of the PDA has to be processed by a powerful computer. The software used by the computer for background correction makes assumptions on the shape of interfering signals often quite remote from the analysis line (*21*). This fact trades off some spectral resolution.

Another system scheduled to be on the market was developed by Cammann et al. (*22*). This again consists of a Beenakker-type cavity (without cooling) mounted on the GC. The light from the discharge is transported to a novel spectrometer by a trifurcated optical fiber bundle. Light from each arm passes through an oscillating interference filter to a separate photomultiplier with lock-in signal amplifier. The lock-in signals should correspond to the background corrected net intensities of the monitored analyte lines. Reported detection limits in the ng range (*22, 23*) and selectivities are not promising.

Finally a plasma emission detector has been developed in our department in cooperation with Anton Paar Company, Graz, Austria (*24, 25*).

ESD Overview

Apart from the GC, the system (Figure 1) consists of the SCP generator, the torch, the fiber-optic link and the ESD-4 main unit. Generator and torch form a compact unit which is mounted directly on top of the gas chromatograph and powered by an external d.c. supply. The main unit contains a four-channel spectrometer based on interference filters, and specially developed fast and low-noise signal conditioning electronics which provide analog output readily handled by a commercial chromatography data station.

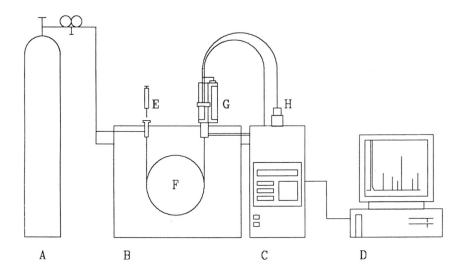

Figure 1.   ESD-4 SYSTEM SETUP

             A ... helium as carrier and plasma gas

             B ... GC oven

             C ... ESD-4

             D ... PC/AT data acquisition board

             E ... injector

             F.... chromatographic column

             G ... SCP

             H ... optical fiber and interference filter spectrometer

Stabilized Capacitive Plasma (SCP). The circuitry of the RF generator is patented (*26 - 28*) and consists of a 27.12 MHz Oscillator, two semiconductor amplification stages and an impedance matching network. This network is formed by a symmetrically tapped inductance and a variable capacitor in parallel to the plasma. Given a plasma gas flowing in an inert tube, two principal configurations are possible. The electric field can be applied in a direction perpendicular to the tube axis, or in a direction parallel to it. Both configurations were explored by using various electrode arrangements. They were evaluated and compared with respect to energy density obtainable in the plasma and interaction of plasma with tube wall and sample. The longitudinal arrangement has proven to be superior (unpublished results, manuscript in preparation). In this arrangement the power dissipated in the discharge was about 140 W, determined from the carefully measured heat uptake of the cooling water and plasma gas. The power input to the SCP was 280 W at 48 V d.c., indicating a 50% efficiency of the generator. The stability of the generator was assessed by monitoring plasma atomic emission of helium at 728.1 nm. Signal stability was better than 0.2% within one hour and better than 1% within 20 hours. Figure 2 depicts the longitudinal electrode arrangement from top and side on.

Torch Design. Around the discharge tube, there are two annular electrodes capacitively coupling energy into the discharge region, which has an overall length of 20 mm. The film thickness of the cooling water passing between the discharge tube and the electrodes is 0.4 mm and the flow rate is 500 ml/min. The capillary column transporting the GC effluent ends approximately 5 mm below the discharge region. The plasma gas used is helium at a flow rate of 10 to 100 ml/min.

Spectral Range. The design of a spectrometer is dominated by the requirements of the spectral region for which it is intended. Emission spectra of nonmetals were investigated with a scanning monochromator (Table I). The compounds chlorobromobenzene and octanethiol were evaporated into the discharge from a thermostated diffusion capillary. The rate of diffusion of a few ng/s was verified gravimetrically. After constant signal levels were attained emission spectra were recorded under a variety of conditions in the range from 650 to 1050 nm (*29*). Results indicate, that many of them have their most intense lines in the NIR. Figures 3 and 4 display the spectral windows around the major emission lines of chlorine and sulfur. It can be seen that even a moderate spectral bandpass will suffice to measure these elements selectively.

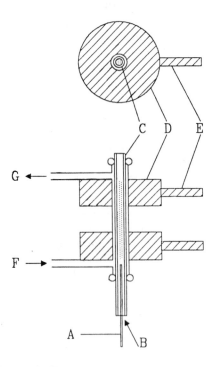

Figure 2. TORCH DESIGN

    A...chromatographic capillary column

    B...plasma gas (helium)

    C...fused silica discharge tube (ID 0.53 mm, OD 0.7 mm)

    D...annular electrode (gold plated brass)

    E...RF-connector

    F...cooling water in

    G...cooling water out

Interference Filter Spectrometer. The simultaneous measurement of intensities at different positions of such a spectrum is usually done with a polychromator. When the requirements on spectral band pass are not excessive, interference-filters can be used to achieve high optical throughput and a compact design. The use of several filters for the simultaneous measurement of multiple elements creates problems.

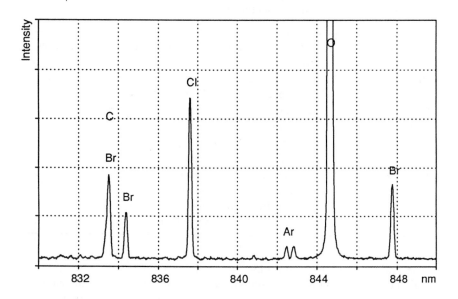

Figure 3. Spectral window around chlorine emission line **837.6** nm, spectroscopic conditions according to table I.

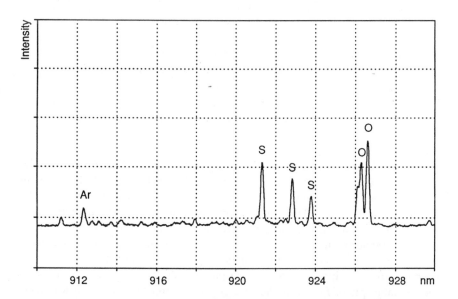

Figure 4. Spectral window around sulfur emission line **921.3** nm, spectroscopic conditions according to table I.

Table I. Spectroscopic equipment

monochromator	HR 640 Jobin-Yvon, Czerny-Turner configuration, focal length 640 mm
grating	holographic grating, 1200 lines/mm, optimized range 600 - 1300 nm
entrance slit	100 um
exit slit	100 um
PMT	R 928 Hamamatsu
photodiode	S2387-16R Hamamatsu
preamplifier units	custom made by Anton Paar Company
spectrometer control	Spectra-Link, Jobin-Yvon
optical fiber	fused silica monofiber, 1 mm diameter, 1.3 m length, Gigahertz Optic, FRG

The optical problem is to divide the available light evenly among different filters and the economic problem is, that high resolution interference filters are expensive and therefore the cost per measurement channel is relatively high. In this context it is important to recall that the passband of interference filters shifts with the angle of incident light. This effect is well known and is suggested by filter manufacturers to fine tune the transmission wavelength of their filters. This effect has also been used by Müller and Cammann (22) in the design of their wavelength modulated plasma detector. In our case this offers the possibility to shift the passband of the interference filter to a wavelength useful for spectral background correction. The "interference filter spectrometer" (IFS) described subsequently, offers a solution to both of the problems mentioned above (patent pending). It manages the even distribution of the available light among several filters and it uses a single filter for the simultaneous measurement of signal and background intensities at neighbouring wavelengths.

The IFS (Figure 5) consists of two sub-units: the collimator and the filter-detector-system. The collimator creates a hollow cylinder of light travelling down to the filter-detector-system by means of a parabolic mirror (f = 80 mm, d = 30 mm). The system contains four prealigned detector units. Each unit consists of a concentric support, an interference filter and two rotatable carrier plates. On each plate a deflecting mirror at 45 degrees to the optical axis sends part of the collimated light horizontally through the filter (two cavities, bandpass = 0.6 nm) to a photodiode with preamplifier mounted on the same plate. Figure 6 shows a schematic top-view of such a detector unit: A cross-section of the hollow cylinder of collimated light, the interference filter in its center, two carrier plates and two deflecting mirrors. Each

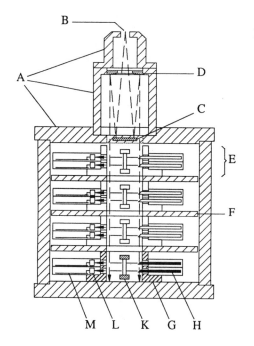

Figure 5.   INTERFERENCE
FILTER SPECTRO-
METER (IFS)

A... aluminum housing

B... end of optical fiber

C... flat mirror

D... concave mirror with
centered hole

E... prealigned detector
unit

F... aluminum plate

G... concentric support

H... deflecting mirror

K... interference filter

L... photodiode

M.. preamplifier board

mirror sends a fraction of the light through the filter, the rest is available for other mirrors further down the instrument. The angle of incidence of this still collimated beam on the filter can be adjusted by simply rotating the carrier plates around the filter holder. A typical value for the background correction position is 1 nm off the analytical wavelength. The arrangement on a carrier plate makes sure that each photodiode always is illuminated by its corresponding mirror.

If the point light source illuminating the collimator is an optical fiber, the light from even an inhomogeneous source is evenly distributed along concentric circles and the different measurement channels are evenly illuminated. This illumination with "equivalent" light is a prerequisite for efficient background correction, and efficient background correction is a prerequisite for selective determination of different elements.

Signal Conditioning. The ESD-4 box contains purpose-built multichannel amplifiers. Data acquisition parameters such as gain, offset and, if required, analog subtraction of the background signal, can be set separately for each channel via a numeric keyboard. The values can be stored in internal memory and recalled and displayed on an alphanumeric display. One signal channel can be selected for continuous display on the front panel.

All the amplified and corrected signals are present as analog outputs, ready for processing by any chromatographic data acquisition system. The data shown in the chromatographic section were acquired with Chromstar (Version 2.07) from Bruker-Franzen Analytik GmbH, Bremen. This is a four channel data acquisition package with 22 bit A/D-converters for signal levels up to 5 V running under Microsoft Windows on a 386-PC from IBM or compatible machines. The standard acquisition rate for recording chromatograms was five data points per second.

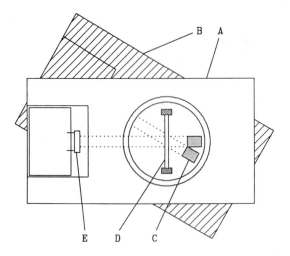

Figure 6.   PREALIGNED DUAL DETECTION UNITS

A ... signal channel detection unit

B ... background detection unit

C ... 45° deflection mirror

D ... narrow-band interference filter

E ... photodiode with signal preamplification

Chromatographic Section.

For the assessment of sensitivity and selectivity a test mixture of 13 different compounds containing the heteroelements chlorine, bromine, fluorine, iodine and sulfur was prepared. A stock solution was obtained by diluting 0.1 mL of each compound in 100 mL dichloromethane. Due to different densities the actual concentration of the compounds varies from 0.84 to 2.96 g/L (Table II). This stock solution was stored at -18°C. All compounds were of analytical grade and purchased from E.Merck, Darmstadt, FRG. For the gas chromatographic separation the stock solution was diluted in n-pentane up to a factor of hundred (8.4 - 29.6 mg/L).

Table II. Composition of the Synthetic Test Mixture

No.	Compound	Formula	Concentration
1	1,1,1-Trichloro-3,3,3-trifluoroacetone	$C_3Cl_3F_3O$	1.50 g/L
2	Tetrachloromethane	CCl_4	1.59 g/L
3	Trichloroethene	C_2HCl_3	1.46 g/L
4	Fluorotoluene	C_7H_7F	1.00 g/L
5	Tetrachloroethene	C_2Cl_4	1.62 g/L
6	Bromobenzene	C_6H_5Br	1.50 g/L
7	1-Bromo-2-fluorobenzene	C_6H_4BrF	1.61 g/L
8	1,2-Dichlorobenzene	$C_6H_4Cl_2$	1.30 g/L
9	Thioanisole	C_7H_8S	1.06 g/L
10	1-Bromo-2-chlorobenzene	C_6H_4BrCl	1.66 g/L
11	1-Octanethiol	$C_8H_{18}S$	0.84 g/L
12	1-Iodooctane	$C_8H_{17}I$	1.32 g/L
13	Tetrabromoethane	$C_2H_2Br_4$	2.96 g/L

Results and Discussion

Sensitivity and Selectivity. Figure 7 shows the separation of the mixture on a 10 m OV-1 column, simultaneously recorded on four different emission wavelengths. Aliquots of 1 uL of a 1:100 diluted stock solution were injected with a split ratio of 1:17. The total amounts vary from 0.49 ng for octanethiol up to 1.74 ng for tetrabromoethane. The solvent peak appearing in the chlorine channel derives from the dichloromethane from the stock solution.

One of the benefits of element specific detection is the possibility of the identification of overlapped peaks when they contain different heteroelements. Figure 8 shows the separation of the same test mixture as in Figure 7 but using a modified temperature program (Table III). On the carbon channel there are three peaks, the last one appearing to be a little bit broader, suggesting an unresolved pair of compounds. An exact comparison of the retention times of the signals in the chlorine, bromine and the sulfur channel shows that there is a time difference of 1.5 seconds between the peak maxima of chlorine and bromine in comparison to sulfur. The two chromato-

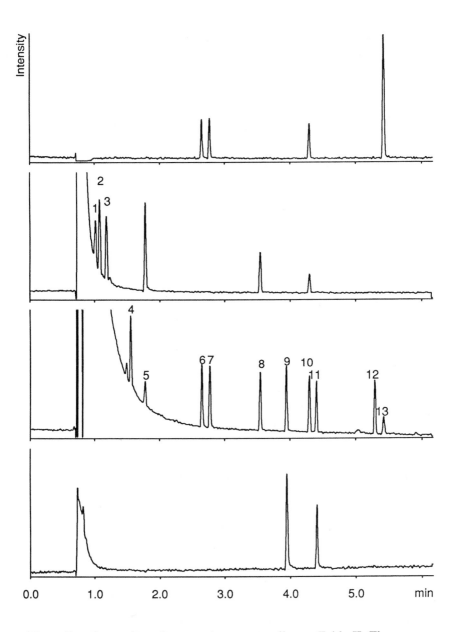

Figure 7.   Separation of a test mixture according to Table II. The
channels are from top to bottom: bromine, chlorine,
carbon and sulfur. Chromatographic conditions according
to Table III.

graphically non resolved peaks are 1-bromo-2-chlorobenzene and octanethiol but clearly in reversed order compared with Figure 7.

Table III. Chromatographic Conditions

Figure Number		7	8	9
Gas Chromatograph		--- Carlo Erba 5160 Mega Series ---		
Injector Temperature	[°C]	200	200	240
Detector		-------------- ESD-4 -----------------		
Detector Temperature	[°C]	300	300	300
Initial Temperature	[°C]	60	100	60
Initial Time	[min]	1	1	2
Temperature Ramp	[°/min]	20	20	10
Final Temperature	[°C]	160	240	250
Final Time	[min]	1	1	2
Column		a	b	a
Column Head Pressure	[kPa]	40	85	45
Injection Mode		split[c]	split[c]	splitless
Injected Amount	[μL]	1[d]	0.5[d]	1[d]

[a] 10 m OV-1 Macherey-Nagel, 0.2 mm i.d., 0.25 μm film thickness
[b] 25 m HP-1 Hewlett-Packard, 0.2 mm i.d., 0.33 μm film thickness
[c] split ratio 1:17
[d] Hamilton #75, 5 μL volume with fixed needle

Application to Industrial Quality Control. The separation of a real sample is shown in Figure 9, displaying the multielement chromatogram of an extracted cardbord sample. For the sample preparation 3 g of cardboard packing material was cut into pieces of approximately 3 by 3 mm. After extraction in a Soxhlet apparatus for ten hours the solvent was evaporated under a stream of dry nitrogen and the residue was dissolved in 0.5 mL of diethyl ether. Aliquots of 0.5 uL were injected splitless into the GC.

The carbon trace gives a universal signal comparable to the signal of a flame ionisation detector, with the added advantage that the signal is proportional to the amount of carbon present in the peak. The great number of unresolved peaks is reduced to few on the sulfur and chlorine channels, while no bromine containing compounds have been detected.

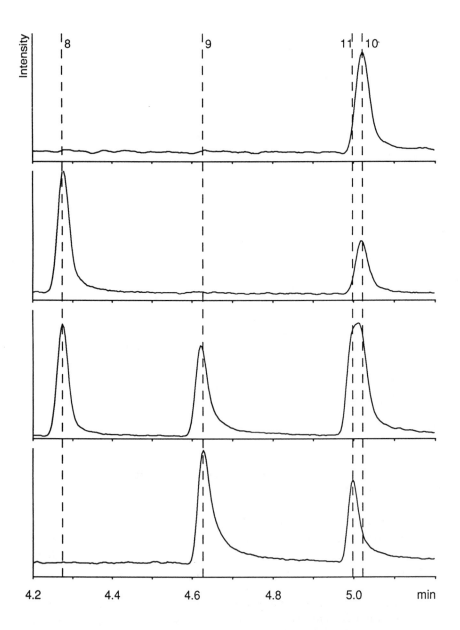

Figure 8.   Overlapping peaks containing different heteroelements. The channels are from top to bottom: bromine, chlorine, carbon and sulfur. Chromatographic conditions according to Table III.

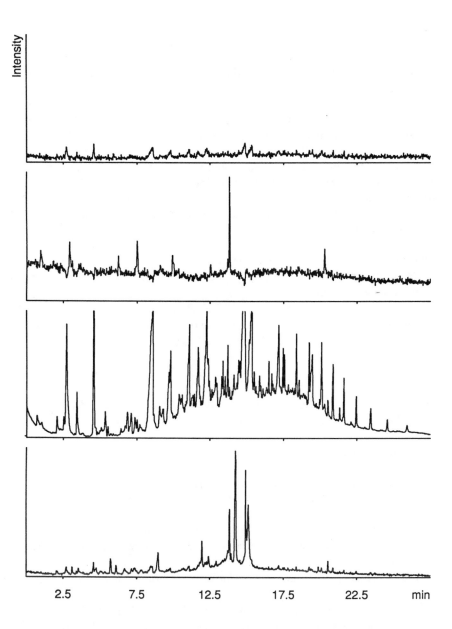

Figure 9.   Chromatogram of an extracted cardboard sample. The
channels are from top to bottom: bromine, sulfur, carbon
and chlorine. Chromatographic conditions according to
Table III.

Conclusion

A complete, modular system is described for element specific detection in gas chromatography consisting of the SCP, a novel plasma source, and its interface to a GC, the optics gathering and transferring the emitted light, a four-channel spectrometer with simultaneous background correction, low-noise electronics for multichannel signal amplification and a PC with hard and software for chromatographic data acquisition.

Powerful features are its compact and rugged design, torch and interface designed for high-temperature GC up to 450°C and simple handling for users without spectroscopic experience. Analog signals at the output terminals of the ESD-4 are compatible with all common integrators and PC-based chromatography data systems.

And finally the most important advantages:
– detection limits in the low [pg/s] range for the elements carbon, chlorine, bromine and sulfur
– truly simultaneous measurement of four elements without compromises in sensitivity. This can save a factor of two or three in analysis time and in addition increases the reliability of data.

Acknowledgements

This work was supported by the "Fonds zur Förderung der Wissenschaftlichen Forschung", Austria, under Grant No. P 7110-CHE.

Literature Cited

(1) Uden, P.C. *Tr. Anal. Chem.* **1987**, *6*, 238-246
(2) McAteer, P.J.; Ryerson, T.B.; Argentine, M.D.; Ware, M.L.; Rice, G.W. *Appl. Spectrosc.* **1988**, *42*, 586-588
(3) Hooker, D.B.; DeZwaan, J. *Anal. Chem.* **1989**, *61*, 2207-2211
(4) Hagen, D.F.; Haddad, L.C.; Marhevka, J.S. *Spectrochim. Acta* **1987**, *42B, 253-267*
(5) McCormack, A.J.; Tong, S.C.; Cooke, W.D. *Anal. Chem.* **1965**, *37*, 1470-1476
(6) Beenakker, C.I.M. *Spectrochim. Acta* **1976**, *31B*, 483-486
(7) Bollo-Kamara, A.; Codding, E.G. *Spectrochim. Acta* **1981**, *36B*, 973
(8) Sobering, G.S.; Bayley, T.D.; Farrar, T.D. *Appl. Spectrosc.* **1988**, *42*, 1023-1025
(9) Fielden, P.R.; Jiang, M.; Snook, R.D. *Appl. Spectrosc.* **1989**, *43*, 1444-1449

(10) Bruce, M. L.; Workman, J.M.; Caruso, J.A.; Lahti, D.J. *Appl. Spectrosc.* **1985**, *39*, 345
(11) Patel, B.M.; Heithmar, E.; Winefordner, J.D. *Anal. Chem.* **1987**, *59*, 2374-2377
(12) Rice, G.W.; D'Silva, A.P.; Fassel, V.A. *Spectrochim. Acta* **1985**, *40B*, 1573-1584
(13) Rice, G.W.; D'Silva, A.P.; Fassel, V.A. *Anal. Chim. Acta* **1984**, *166*, 27-38
(14) Chan, S.-K.; Montaser, A. *Spectrochim. Acta* **1987**, *42B*, 591-597
(15) Houk, R.S.; Fassel, V.A.; Lafreniere, B.R. *Appl. Spectrosc.* **1986**, *40*, 94-100
(16) Freeman, J.E.; Hieftje, G.M. *Appl. Spectrosc.* **1985**, *39*, 211
(17) Freeman, J.E.; Hieftje, G.M. *Spectrochim. Acta* **1985**, *40B*, 475-491
(18) Pivonka, D.E.; Schleisman, A.J.J.; Fateley, W.G; Fry, R.C. *Appl. Spectrosc.* **1986**, *40*, 766-772
(19) Hubert, J.; van Traa, H.; Tran, K.C.; Baudais, F.L *Appl. Spectrosc.* **1986**, *40*, 759-766
(20) Quimby, B. D.; Sullivan, J. J. *Anal. Chem.* **1990**, *62*, 1027-1034
(21) Sullivan, J. J.; Quimby, B. D. *Anal. Chem.* **1990**, *62*, 1034-1043
(22) Müller, H.; Cammann, K. *J. Anal. Atom. Spect.* **1988**, *3*, 907-913
(23) Stilkenboehmer, P.; Cammann, K. *Fresenius' Z. Anal. Chem.* **1989**, *335*, 764-768
(24) Knapp, G.; Leitner, E.; Michaelis, M.; Platzer, B.; Schalk, A. *Intern. J. Environ. Anal. Chem.* **1990**, *38*, 369-378
(25) Gross, R.; Leitner, E.; Platzer, B.; Schalk, A.; Knapp, G.; Grillo, A. *Pittsburgh Conference 1990*, Abstract No. 1295
(26) Knapp, G.; Schalk, A. Ger. Pat. PE 36 38 889 A1 (1987)
(27) Knapp, G.; Schalk, A. United States Pat. 4,877,999 (1989)
(28) Knapp, G.; Schalk, A. UK Pat. GB 2 183 087 B (1990)
(29) Gross, R. diploma thesis, Graz Technical University, 1989

RECEIVED April 12, 1991

Chapter 10

Alternating-Current Plasma Detection for Gas Chromatography and High-Performance Liquid Chromatography

E. F. Barry, L. A. Colon, and R. B. Costanzo[1]

Department of Chemistry, University of Lowell, Lowell, MA 01854

The alternating current plasma (ACP) for GC and HPLC offers considerable potential for selective element detection. The detector is simple in design and produces a stable self-seeding discharge which does not extinguish with high solvent concentration. The device has been used for the gas chromatographic determination of organomercury, lead and chlorine and the determination of organomercury by HPLC.

Atomic Emission Spectroscopy (AES), in tandem with chromatographic techniques, has become quite popular in the last decade to facilitate interpretation of complex chromatograms through element-specific detection. Plasma Emission Detectors (PED) have enjoyed popularity in GC and HPLC because their characteristics are attractive for speciating purposes. The desirable attributes of element-specific detectors have been reported (1-6), high sensitivity and selectivity, in addition to the possibility of simultaneous multielement determinations, being among their features.

Plasma Emission Sources

There are three principle plasma emission sources which have been successfully interfaced with both GC and HPLC: Inductively Coupled Plasma (ICP), Direct Current Plasma (DCP) and Microwave-Induced Plasma (MIP). The interfacing of MIP with HPLC has been more difficult due to its low tolerance to typical flow rates of organic solvents. In the following sections the development and performance of an alternative PED for GC and HPLC, the alternating current plasma (ACP), will be described.

[1]Current address: Pharmaceutical Development Laboratory, Astra Pharmaceutical Products Company, 50 Otis Street, Westborough, MA 01581

0097–6156/92/0479–0170$06.00/0
© 1992 American Chemical Society

Alternating Current Arcs. Alternating Current (ac) arcs have long been employed in quantitative detection and determination of trace elements (*7*). Typically ac arcs are sustained at high voltage at a current of about 2-5 amperes for the analysis of solid and liquid samples. For solid samples of conducting materials, the sample is used as an electrode after being machined to an appropriate form. For nonconducting materials the sample is mixed with carbon powder and packed into an electrode cavity. Liquid samples are commonly deposited into a miniature cup which has been machined on one of the electrodes. The sample is exposed to the discharge for a period of time, usually 30 seconds, and the intensity of emitted radiation is measured. The electrodes in an ac discharge heat less than in a DC arc; thus, electrode corrosion is minimal. It is also well known that ac arcs are more stable than the dc versions and provide better reproducibility; however, the sensitivity is decreased. Plasma spectroscopic studies on ac arcs in methane have also been investigated (*8*); nevertheless, no attempts have been made to couple them to chromatographic techniques.

A microarc was developed as a sample introduction device for the MIP (*9*). It was reported later that the microarc operating with helium may be used as an emission source for atomic spectroscopy (*10*) and also was suggested as a detector for GC (*11*). The helium species in the plasma yield sufficient collisional energy transfer to excite other elements and produce characteristic elemental emission. The microarc incorporates a high voltage, low current full- or half-wave rectified and unfiltered power supply to generate the ac discharge.

Alternating Current Plasma. The ACP operates basically by the same principle as the microarc; however, the high voltage is not rectified, maintaining the same frequency of the power line. It is probably the simplest and least expensive atomic source to construct and operate. A uniform discharge is generated across two electrodes in a controlled helium atmosphere. A primary supply of 115 V ac at a frequency of 60 Hz feeds a step-up transformer which provides an ultimate output from its secondary of approximately 10,000 V at a current of 23 mA. The polarity of the arc reverses at a frequency of 120 Hz when a 60 Hz source is employed. The discharge is confined into a quartz discharge tube with the emitted radiation focused on the entrance slit of a monochromator by means of biconvex lens, then detected and processed. The ACP acts as a stable, self-feeding plasma such that external initiation is not required.

GC-ACP DETECTION

Since the ACP is operated at atmospheric pressure, the eluent of a capillary column is easily directed towards the discharge tube containing the plasma. The configuration of a typical discharge tube interface is illustrated in Figure 1. Helium make-up gas is introduced into the plasma to stabilize the discharge from disturbances that might arise when the eluent sample peaks from the

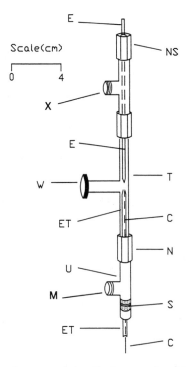

Figure 1. Schematic diagram of the discharge tube: (C) megabore capillary column, (S) silicon septa-electrode holder, (M) helium make-up flow, (ET) bottom electrode, (N) nut and ferrules, (W) quartz window, (T) discharge tube, (X) exhaust, (U) union tee, (NS) nut and septum, (E) top electrode. (Reprinted with permission from Ref. *12, Applied Spectroscopy*, Copyright 1988, The Society for Applied Spectroscopy.)

chromatographic column reach it. The make-up gas also provides a stable plasma; this is particularly important at low column flow-rates. The "T" configuration allows a transverse view of the plasma through the quartz window which is mounted on the end of the window arm and forms an airtight tube. In this configuration degradation products produced by the plasma are carried away from the window arm, avoiding any blockage of the emitted radiation by deposits which occurs when straight discharge tubes are employed. Accumulation of deposits changes discharge tube properties with time and leads to loss in sensitivity. The airtight electrode holders provide a means of confining the discharge to a helium atmosphere and minimize spectral interferences.

GC-ACP Instrumentation. A schematic representation of the GC-ACP arrangement is depicted in Figure 2 and a description of the various components and operating conditions is presented in Table I. The operating conditions of the GC-ACP system have been established by simplex optimization (*12*). The fused silica capillary column is extended from the column oven to the center of the bottom electrode by means of a flexible metal transfer tube which is heated to the required temperature. The bottom electrode is electrically insulated from the transfer tube by means of a rubber septum. The column, which is guided through the flexible metal tube, is inserted through a copper electrode tube (1/8 inch o.d.) machined to a point of approximately 0.5 mm in diameter (see Figure 1). The column end is positioned approximately 0.5 mm from the electrode tip. This design minimizes losses in sensitivity due to extra-column dispersion that would produce excessive band broadening.

An analytical emission line to be monitored is selected after purging a helium stream saturated with vapor of a chemical species containing the element of interest and scanning the appropriate wavelength regions. Impurities in helium such as water, hydrocarbons, nitrogen and oxygen that are not removed by traps contribute to the plasma background emission. The emission profiles of tetrabutyllead and chloroform were used to establish the analytical lines of Pb(I) 283.4 nm and of C-Cl 278.82 nm for the GC-ACP of lead and chlorine, as described in the following sections.

Organomercury Detection. In our initial study of the ACP (*13*), methyl- and ethyl- mercury(II) chlorides were used as probe solutes. The detection limits of the ACP for these organomercurial species were 3.5 and 20 pg/s (as mercury), calculated by the expression given by Scott (*14*) and Braman and Dynako (*15*). The detection limits were improved to 1.1 pg/s after simplex optimization of the GC-ACP (*12*), comparing favorably with the detection limits of the GC-MIP (*16,17*). The linear dynamic range was three orders of magnitude for both solutes.

Mercury selectivity was evaluated in a mixture containing carbon as n-octane, mercury as methylmercury(II) chloride, oxygen as 1,4-dioxane, nitrogen as di-n-propylamine and chlorine as carbon tetrachloride. These elements were

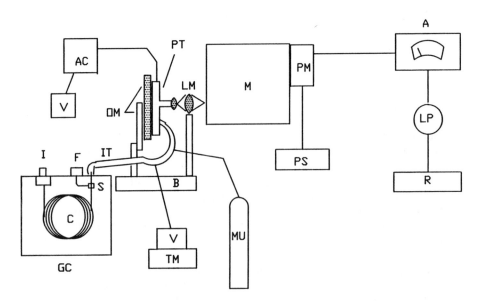

Figure 2. Schematic diagram of the GC-ACP experimental arrangement: (GC) gas chromatograph, (C) capillary column, (I) injection port, (F) FID, (S) column oven tee split, (IT) detector interface tube, (V) variac, (AC) ac power supply, (OM) optical bench, (PT) plasma discharge tube, (LM) lens and mount, (MU) helium make-up gas, (B) optical bench, (TM) thermocouple thermometer, (M) monochromator, (PM) photomultiplier tube, (PS) PMT power supply, (A) picoammeter, (LP) low-pass filter, (R) recorder integrator. (Reprinted with permission from Ref. *22, Journal of Chromatography*, Copyright 1989, Elsevier Science Publishers B.V.)

Table I. Instrumental components and general operating conditions

Component	Experimental Condition
Gas Chromatograph Hewlett-Packard 5890A Columns: 30m x 530μm, thickness: 3μm DB-1 : 10m x 530μm, thickness: 4.8μm CP SIL8 CB	injector temperature: 220 °C interface temperature: 200 °C oven temperature: 180 °C column flow-rate: 20 mL/min injection volume: 0.5-1 μL split ratio: 10/1 to 16/1 make-up flow-rate: 1-30 mL/min
Liquid Chromatograph Pump: Spectra Physics SP8700 Injector: Rheodyne 7125 Column: 4.6 x 250 mm ODS, 5μm	mobile phase: MeOH/H_2O (20/80) injection volume: 25 μL
Copper electrodes 1-3mm o.d.	gap: 4-15mm
ac power supply: Furnace ignition transformer (France, Fairview, TN)	Maximum output: 11,200 V ac
Monochromator: McPherson EU-700 (McPherson, Acton, MA)	slit widths: 50-1200 μm slit height: 5 mm
Photomultiplier tube: Hamamatsu, model R212 (Middlesex, NJ)	voltage: -1000V
Lens: Oriel (Stratford, CT)	75 mm biconvex quartz
Picoammeter: Keithley Instruments (cleveland. OH)	range: 0.03-3 x 10^{-6}Amps
RC low-pass filter	time constant: 0.2 s
Data acquisition system (Galactic Industries, Salem, NH)	Chrom-1AT and Lab Cal software

selected because they have characteristic emission lines near the mercury line of 253.67 nm. The selectivity of the detector was defined as the ratio of the ACP response of Hg I at 253.67 nm per unit mass of mercury injected to the ACP response toward O, N, C or Cl at 253.67 nm per unit mass of that element injected. Selectivities are given in Table II. It should be noted that ACP selectivity toward Hg I vs. carbon as n-octane is approximately five times

Table II. Selectivity of the GC-ACP towards ethyl and methylmercury Chloride at 253.67 nm, as compared to carbon, chlorine, oxygen, or nitrogen.

Compound	Element Compared	Selectivity Ratio
n-Octane	Carbon	530,000
1,4-Dioxane	Oxygen	300,000
Di-n-propylamine	Nitrogen	84,000
Carbon tetrachloride	Chlorine	30,000

greater than the corresponding selectivity of a MIP detector for Hg I towards carbon as dodecane (16). On the other hand, the selectivity for Hg I towards Cl was comparatively low; this may be attributed to spectral interference caused by C-Cl species formed in the plasma under the power conditions utilized.

In Figure 3 a chromatogram of a gasoline sample spiked with organomercurials is displayed, peaks A and B representing 4.2 and 1.8 ng of methylmercury(II) chloride and ethylmercury(II) chloride, respectively. At the beginning of the chromatogram many hydrocarbons elute causing a slight cooling of the plasma and a noisy baseline, in addition to a large proportion of molecular emission. Even though the background matrix has a high organic concentration, the plasma does not extinguish because of its inherent self-seeding nature; thus, the need for solvent venting is eliminated.

Organolead Detection. The role of ACP for the determination of organolead has also been studied (18) because of the continued concern for the impact of release of lead compounds into the environment. Tetrabutyllead (TBL) and a commercial mix containing tetraethyllead (TEL), ethylene dibromide and ethylene dichloride were chosen as evaluating solutes. The detection limit for TBL at 283.42 nm was established as 130 pg/s which compares quite favorably with the limits determined by GC/DCP, HFID and GC/AA for TBL and other lead compounds (19-21). The linear dynamic range for TBL was 3.5 orders of magnitude. The detection limit for TEL is slightly better because of more rapid elution and sharper peak profile. The Pb/C selectivity of the ACP at 283 nm was 13,800, determined by the procedure described previously for mercury selectivity, except that TBL and n-

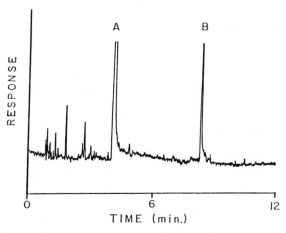

Figure 3.   Gas chromatogram of a mixture of gasoline, spiked with 92.8 ng/μL methylmercury(II) chloride and 39.6 ng/μL ethylmercury(II) chloride. The separation was done at an oven temperature of 110°C and a helium flow-rate of 6 mL/min.   Peak A represents 4.18 ng of methylmercury (II) chloride and peak B represents 1.77 ng of ethylmercury(II) chloride. (Reprinted with permission from Ref. *13*, *Analytical Chemistry*, Copyright 1988, American Chemical Society.)

dodecane were used as analytes. Even without background correction this compares favorably with the reported selectivities using GC/DCP and GC/MIP (17,21). With the ACP, the lead selectivity at secondary lines such as 368 and 405 nm was poorer due to the contribution of carbon background emission at these wavelengths whereas at 283.42 nm carbon emission accounted for only 2.1 percent of the total emission intensity.

In order to demonstrate the lead selectivity of the ACP several applications were performed again with gasoline as a potentially problematic matrix. Leaded and unleaded gasoline were spiked and chromatographed by GC-ACP. The TEL present in the leaded gasoline along with the spiked TBL and another unidentified component of the gasoline were identified. The rapid elution of hydrocarbons is responsible for the perturbation at the beginning of the chromatograms and results in a brief period of carbon background emission, detector overload and plasma instability.

The ACP also exhibits "tuneable" selectivity; for example, chromatograms of an isooctane sample containing 1000 ppm of TEL and smaller levels of ethylene dichloride and dibromide generated at the appropriate analytical wavelengths appear in Figure 4, along with the FID chromatogram (Figure 4a).

Organochlorine Detection. The response of the ACP to selected organochlorine species has also been studied (22) with the following probe solutes: tetrachloroethylene (1.6), n-butylchloride (1.1), p-dichlorobenzene (1.5) and 1,10-dichlorodecane (0.71) where the numbers in parentheses indicate the detection limit of the species in units of ng/s at 278.84 nm; linearity was three orders of magnitude. Detection limits and selectivity are somewhat affected by the molecular emission of the C-Cl band which is lower in intensity than atomic emission. Because required equipment to monitor the major atomic emission lines of chlorine in the near-IR region was not available (e.g., PMT), and emission from the Cl(II) species was not observed (due to intense background emission), the molecular species C-Cl was studied (Figure 5) (23,24). The maximum signal to noise ratio was obtained with a slit width of 1500 um (bandpass 0.3 nm) but, in order to reduce the inclusion of adjacent diatomic carbon swan bands, the slit width was reduced to 500 um (bandpass 0.1 nm). The background which overlapped with the primary 278.84 nm C-Cl band could not be circumvented without adversely affecting sensitivity because of a substantial decrease in energy throughput. However, at this wavelength the overlap of the C_2 band contributed less than 3.5 percent to the C-Cl intensity; at other wavelengths overlaping accounted for more than 10 percent of the C-Cl intensity.

The selectivity for organochlorine is defined as the ratio of the peak area response of the ACP towards C-Cl (originating from 1,10-dichlorodecane) at 278 nm per gram of C-Cl divided by the peak area response towards diatomic carbon (originating from n-dodecane) per gram of diatomic carbon. The selectivity of 24 for the molecular species C-Cl compares favorably with selectivity data obtained with GC-MIP (24) for the same molecular species. The

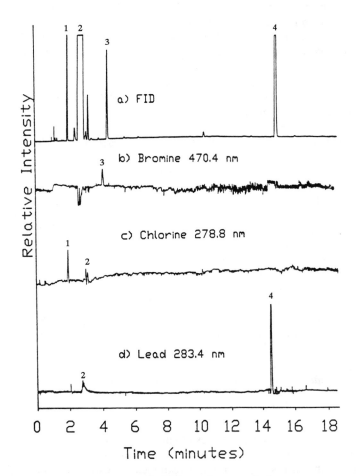

Figure 4. Chromatograms of TEL and organohalogen compounds in isooctane at different wavelengths. Column temperature: 50°C (2 min) to 150°C at 6°C/ min, 1 µL at a split ratio of 10/1. (a) FID chromatogram of isooctane mix at an attenuation of 3, peak (2) represents iso-octane; (b) ACP chromatogram at 470.4 nm: peak (3) represents ethylene dibromide; (c) ACP chromatogram at 278.8 nm: peak (1) represents ethylene dichloride; (d) ACP chromatogram at 283.4 nm: peak (4) represents TEL. (Reprinted with permission from Ref. *18, Journal of High Resolution Chromatography and Chromatography Communications*, Copyright 1989, Dr. Alfred Huethig Publishers.)

effect of molecular structure on C-Cl response was studied by generating relative response factors for a series of compounds having different aliphatic and aromatic character. Using n-butyl chloride (NBC) as an internal standard, the relative response factor is defined as the mass of C-Cl in the analyte of interest per mass of C-Cl in NBC times the ratio of the peak area of NBC to the analyte peak area. Table III clearly shows the emergence of a complex

Table III.   Relative response factors (RRF) for various alkylchlorides and aromatic chlorides[a]

Compound	Number of Carbon in Backbone	Number of Cl Moieties	RRF
1-Chloropropane	3	1	1.21
n-Butylchloride	3	1	1.00
n-Pentylchloride	5	1	1.87
n-Hexylchloride	6	1	2.05
3-Chloroheptane	7	1	2.23
o-Chlorotoluene	7	1	2.56
p-Dichlorobenzene	6	2	3.73
1,6-Dichlorohexane	6	2	2.33
1,2,3-Trichloropropane	3	3	5.61
Tetrachloroethylene	2	4	3.55
1,10-Dichlorodecane	10	2	2.82

[a]Reprinted with permission from Ref. 22, *Journal of Chromatography*, Copyright 1989, Elsevier Science Publishers B.V.

trend. Within the homologous series of chloroalkanes a nonlinear increase in response is probably due to the increase in diatomic carbon background emission with increasing carbon backbone.   Also, for the pair, p-dichlorobenzene vs. 1,6-dichlorohexane (its aliphatic counterpart) the aromatic ring introduces a greater contribution to the response factor than the alkane, reflecting the greater number of diatomic carbon molecules produced per aromatic compound.   The same is true for the pair, o-chlorotoluene and 3-chloroheptane.   A more extensive series of organochlorine compounds would have to be studied in order to thoroughly understand the dependency of response on molecular structure.

Parallel chromatograms with FID and ACP detection of a gasoline sample spiked with lindane are displayed in Figure 6 and clearly illustrate the selectivity of the ACP for detection of organochlorine in a complex matrix.

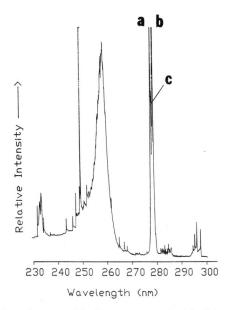

Figure 5. Wavelength scan of helium saturated with chloroform vapor; (a) CCl 277.83nm, (b) CCl 278.84 nm, (c) C₂ interference emission band *ca.* 277.0nm - 278.9nm. (Reprinted with permission from Ref. *22, Journal of Chromatography*, Copyright 1989, Elsevier Science Publishers B.V.)

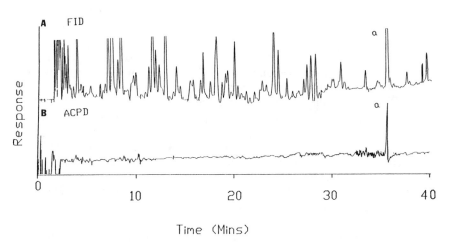

Figure 6. (A) lindane, peak (a) in gasoline detected by FID; GC conditions: 35°C (5 min) to 100°C at 4°C/min then to 230°C at 5 °C/min on DB-1 megabore column. (B) lindane in gasoline detected by ACP, same GC conditions as above. (Reprinted with permission from Ref. *22, Journal of Chromatography*, Copyright 1989, Elsevier Science Publishers B.V.)

HPLC-ACP Detection

Interfacing an ACP with HPLC has proved more challenging than with GC primarily due to the presence of high organic solvent concentration in the mobile phase. The ICP and DCP detectors, can tolerate the introduction of a higher organic concentration than others, e.g., the MIP. In addition, band broadening may be increased by post-column connecting tubing and fittings with any interface. Similar approaches have been used in interfacing DCP and ICP with HPLC, the column eluent being connected to a typical plasma sample introduction device which permits nebulization. However, the sensitivity of such an approach is dependent on the transfer efficiency of the particular nebulizer employed. New sample introduction devices such as the direct injection nebulizer (DIN) (25,26) with a transfer efficiency of nearly 100%, provide a very practical interface.

The low powered helium-MIP, on the other hand, has a very low tolerance towards organic solvents typically used in HPLC, a situation which has been successfully addressed by mixing the plasma gas with oxygen and modifying the discharge tube (27). Another path has been suggested by Zhang et al. (28), who utilized a sample transportation interface based on a moving wheel in conjunction with a heated nitrogen flow for desolvation. The more powerful and costly MIP detectors can tolerate the nebulization of solutions (29,30); however, little attention has been focused on their use as HPLC detectors.

HPLC-ACP Instrumentation. For HPLC-ACP several obstacles must be addressed. First, since the ACP is sustained with helium, conventional nebulizers designed for argon can not be used; the required high pressure nebulizers are not readily available. Secondly, use of argon as the plasma support gas produces a very thin plasma arc which frequently becomes unstable when liquid aerosols are introduced. In addition, the excitation potential of argon is lower than that of helium; thus, a mixture of helium and argon plasma support gas yields the same behavior as observed for argon alone.

We have evaluated several devices for coupling an ACP to a HPLC by means of a glass frit nebulizer (GFN). A promising configuration appears in Figure 7 which illustrates a schematic diagram of a "home-made" GFN; this design closely resembles that used in HPLC-ICP (31). It consists of a modified 15 mL Pyrex sintered glass funnel filter, 20mm in diameter with a frit of 4.5-5.0 μm in porosity. A piece of high pressure tubing (e) connects the column with the GFN. The volume of the chamber is approximately 5mL which can be adjusted by the rubber stopper (d). The upper arm (b) of the modified funnel is attached to the bottom end of the discharge tube, illustrated in Figure 1, by means of a Teflon tee. Helium gas (a) is introduced into the opposite side of the frit as the nebulizing gas and simultaneously supports the ACP. The GFN generates a very fine mist with a droplet size distribution smaller than that from the pneumatic nebulizer, thereby increasing the sample transfer efficiency to the plasma (32), and simultaneously enhancing the

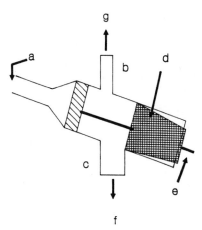

Figure 7. Frit Nebulizer: a) helium flow, b) 4 mm i.d. transfer tube, c) 8 mm i.d. tube, d) rubber stopper, e) HPLC eluent, f) to drain, g) to the ACP.

introduction of organic solvents into the plasma (*33*). However, this fine mist and small inner diameter of the transfer tube (b) have produced condensation of some of the nebulized solution on the inner walls of the transfer tube, which may fill part of the transfer tube with droplets formed and cause disturbances to the plasma. The most easily implemented solution to this problem is to maintain the transfer tube and the connection to the discharge tube above 100°C.

The ACP can be sustained with a flow of approximately several hundred mL/min of helium when an aqueous aerosol is introduced into it, but, higher flow-rates are required to nebulize the column eluent efficiently. The helium flow-rate through the frit may also vary with the analyte under consideration. For example, Figure 8 shows the dependence of mercury (as methyl mercury chloride) response on helium flow-rate. A drawback of the GFN is the eventual clogging of the frit when salt solutions are used in the mobile phase.

Since methanol and acetonitrile are the most commonly used organic modifiers in reverse phase HPLC, the emission spectra of their binary mixtures with water were first obtained. Plasma background spectra between 200 and 450 nm are shown in Figures 9a-9c for aqueous aerosols of water (9a) and mixtures with methanol (9b) and acetonitrile (9c) introduced into the plasma. The background emission consists mainly of OH, NH, N_2, CN, and NO molecular emission, in addition to atomic helium and copper (from the electrodes), the intensity depending on the concentration of methanol or acetonitrile as well as on the quality of the helium used. As discussed previously, an analytical wavelength must be carefully selected in an "open window" of the background emission spectrum of the solvent to avoid possible spectral interference.

The ACP does not extinguish when an aerosol of pure methanol is introduced, but, aqueous mobile phase containing 10% acetonitrile extinguishes the plasma after 5 minutes. This suggests that the ACP with a GFN would be very prone to extinguish if used with normal phase HPLC in which solvents of high carbon content are routinely employed. Ideally, a desolvation system

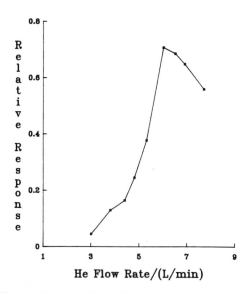

Figure 8. Effects of the helium flow-rate on the ACP response of methylmercury(II) chloride.

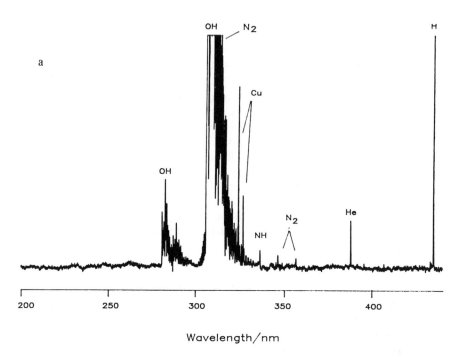

Figure 9. Background emission of the ACP in the region between 200 and 450 nm when aerosols of (a) water, (b) methanol/water (10/20) and (c) acetonitrile/water (5/95) are introduced to the plasma. *Continued on next page.*

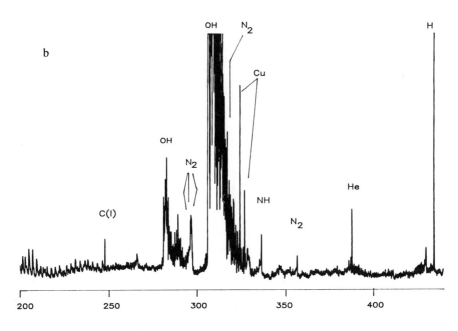

Wavelength/nm

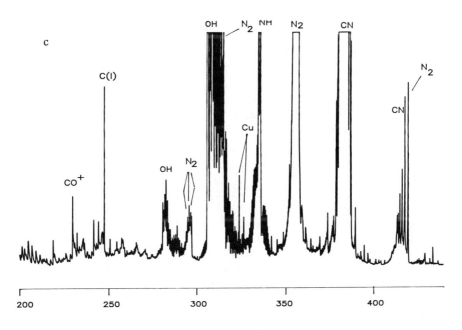

Wavelength/nm

Figure 9. Continued.

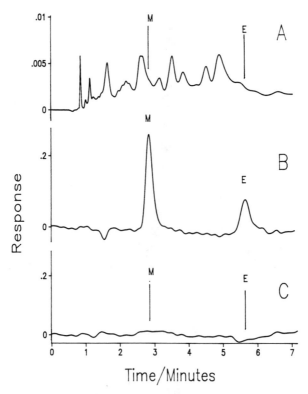

Figure 10. Chromatograms of a gasoline sample, spiked with MMC and EMC, detected by A) UV and B) ACP. C) Unspiked gasoline detected by ACP. M and E refer to MMC and EMC, respectively. Chromatographic conditions: column, Hypersil ODS 5μm (4.6mm i.d. x 100mm); mobile phase, 0.01% 2-mercaptoethanol in methanol/water (22/78); flow-rate, 1.0 mL/min; injection volume of 25 μL.

which effectively removes the organic component of a chromatographic solvent mixture is needed for versatile HPLC-ACP capability.

Organomercury detection. In the evaluation of the ACP as a HPLC detector, methylmercury(II) chloride and ethyl- mercury(II) chloride have been again used as analytes (Colon, L. A. and Barry, E. F. *J. Chromatogr.* **1990**, in press.) At 253.67 nm the detection limits were calculated to be 4.5 and 2.2 ng/s Hg for MMC and EMC, respectively. These values correspond to a 25 μl injection of a solution containing 2.8 ng/μl MMC (2.2 ng Hg/μl) and 4.0 ng/μl EMC (3.0 ng Hg/μl) and represent almost a 100 fold improvement over the detection limits reported for the same solutes by Krull and co-workers (*34*), 232 and 302 ng Hg/μl for MMC and EMC, respectively, determined by HPLC-ICP with a conventional nebulizer.

The selectivity of the ACP has been demonstrated by spiking MMC and EMC to samples of different complex matrices, samples of a local river and gasoline. Chromatograms of each spiked sample were monitored by parallel UV detection at 254 nm and the ACP. In Figure 10 the chromatograms of spiked gasoline are presented; peaks M and E represent 840 and 668 ng of MMC and EMC, respectively. They illustrate the selective detection of the organomercurials with the ACP.

In summary the ACP is a viable alternative to other element-selective detectors, and is easily assembled and interfaced with chromatographic equipment at modest cost. The device exhibits a remarkably stable signal because the plasma is self-seeding and reignites itself every half cycle which is 120 times per second for the 60 Hz power supply used in the work reported here. A tesla coil is not required to commence operation of the plasma if the ac voltage is greater than the breakdown voltage. As a result, the ACP can tolerate large injections of solvent without extinguishing and, thus, requires no venting valve, thereby minimizing band broadening.

The reduction of background spectral interference produced by molecular emission has not yet been investigated with the ACP. Furthermore, the utilization of techniques such as lock-in amplifiers in conjuction with oscillating quartz plates incorporated within a monochromator and pulsed power sources often lead to an improved signal to noise ratio and better selectivity. The carbon mode for selective GC detection has been investigated in our laboratory but detection limits for carbon have been quite high because of the cool temperature of the plasma. In addition, our ACP studies to date have been confined to sequential analyses for multielement detection but with a properly configured interface and a photo- diode array the ACP can be easily converted to a simultaneous multielement selective detector for GC and HPLC.

Literature Cited:
(1)   VanLoon, J. C. *Anal. Chem.* **1979**, *51*, 1139A.
(2)   Jewett, K. L. and Brickman, F. E. *J. Chromatogr. Sci.* **1983**, *23*, 205.
(3)   Ebdon, L.; Hill, S. and Ward, R. W. *Analyst* **1987**, *112*, 1.
(4)   Uden, P. C. In *Developments in Atomic Plasma Spectrochemical Analysis*; Barnes, R. M., Ed.; John Wiley & Sons, 1983, pp.302-320.
(5)   Uden, P. C. *Chromatogr. Forum* **1986**, *1*, 17.
(6)   Uden, P. C. *Trends in Anal. Chem.* **1987**, *6*, 239.
(7)   Pinta, E. *Detection and Determination of Trace Elements*, Ann Arbor Science, Ann Arbor, Michigan, 1966.
(8)   Heinrich, G.; Nikel, M.; Mazurkiewicz, M. and Arvni, R. *Spectrochim. Acta* **1978**, *33B*, 635.
(9)   Layman, L. and Hieftje, G. M. *Anal. Chem.* **1975**, *47*, 194.
(10) Churchwell, M. E.; Beeler, T.; Messman, J. D and Green, R. B. *Spectrosc. Lett.* **1985**, *18*, 679.
(11) Green, R. B. and Williams, R. R. *Anal. Chim. acta*, **1986**, *187*, 301.
(12) Costanzo, R. B. and Barry, E. F. *Appl. Spectrosc.* **1988**, *42*, 1387.
(13) Costanzo, R. B. and Barry, E. F. *Anal. Chem.* **1988**, *60*, 827.

(14) Scott, R. P. W. *J. Chromatogr. Sci.* **1971**, *9*, 645.
(15) Braman, R. S. and Dynako, A. *Anal. Chem.* **1968**, *40*, 95.
(16) Quimby, B. D.; Uden, P. C. and Barnes, R. M. *Anal. Chem.* **1978**, *50*, 2112.
(17) Estes, S. A.; Uden, P. C. and Barnes, R. M. *J. Chromatogr.* **1982**, *239*, 181.
(18) Costanzo, R. B. and Barry, E. F. *J. High Resolut. Chromatogr. Chromatogr. Commun.* **1989**, *12*, 522.
(19) Radojevic, M.; Allen, A.; Rapsomankis, S. and Harrison, R. M. *Anal. Chem.* **1986**, *58*, 658.
(20) Hill, H. H. and Aue, W. A. *J. Chromatogr.* **1976**, *122*, 515.
(21) Poirier, C. A. *Ph.D. Dissertation*, University of Massachusetts, 1982.
(22) Costanzo, R. B. and Barry, E. F. *J. Chromatogr.* **1989**, *467*, 373.
(23) Dagnall, R.; West, T. and Whitehead, P. *Anal. Chim. Acta* **1972**, *60*, 31.
(24) McCormack, A.; Tong, S. and Cooke, W. *Anal. Chem.* **1965**, *37*, 1470.
(25) Lawrence, K. E; Rice G. W. and Fassel, V. A. *Anal. Chem.* **1984**, *56*, 289.
(26) LaFreniere, K. E.; Fassel, V. A. and Eckels, D. E. *Anal. Chem.* **1987**, *59*, 879.
(27) Kollotzeek, D.; Oechsle, D.; Kaiser, G.; Tschopel, P. and Tolg, G. *Fresenius' Z. Anal. Chem.* **1984**, *318*, 485.
(28) Zhang, L.; Carnahan, J. W.; Winans, R. E. and Neil, P. H. *Anal. Chem.* **1989**, *61*, 895.
(29) Hass, D. L.; Carnahan, J. W. and Caruso, J. A. *Appl. Spectrosc.* **1983**, *37*, 82.
(30) Michlewicz, K. M. and Carnahan, J. W. *Anal. Lett.* **1987**, *20*, 193.
(31) Ibrahim, M.; Nisamaneepong, W.; Hass, D. L. and Caruso, J. A. *Spectrochim. acta* **1985**, *40B*, 367.
(32) Layman, L. R. and Lichte, F. E. *Anal. Chem.* **1982**, *54*, 638.
(33) Nisamaneepong, W.; Hass, D. L. and Caruso, J. A. *Spectrochim. Acta*, **1985**, *40B*, 3.
(34) Krull, I. S.; Bushee, D. S.; Schleicher, R. G. and Smith, S. B., Jr. *Analyst* **1986**, *11*, 345.

RECEIVED April 11, 1991

Chapter 11

Helium Surface-Wave Plasmas as Atomic Emission Detectors in Gas Chromatography

S. Coulombe[1], K. C. Tran, and J. Hubert

Department of Chemistry, Université de Montréal, P.O. Box 6210, Station A, Montréal, Quebec, Canada H3C 3J7

The helium surface wave induced plasma is a very attractive atomic emission detector in gas chromatography. The plasma physical characteristics (excitation temperature, gas temperature and electron density) are presented. The emission spectra of non-metals in the UV-visible are described and the effect of the plasma operating parameters on the atomic emission intensity are reported. Applications of the helium surface wave induced plasma in gas chromatography are also presented.

Over the past twenty years, microwave induced plasmas (MIP) have been used as excitation sources for atomic emission spectroscopy. Their major application in analytical chemistry has been in gas chromatography (1-4), since the low power associated with the MIP generation and its operation under reduced pressure does not allow an easy liquid or solid sample introduction. However, the MIP can be sustained in a larger variety of plasma gases than for the inductively coupled plasma (ICP) and therefore an efficient excitation of more elements can be expected.
The first devices used to sustain MIPs could only generate helium plasma at low pressures (5). In 1977, Beenakker (6) described a new TM$_{010}$ cavity, which could be used to operate a helium plasma at atmospheric pressure and since then this device has been used as a detector in gas chromatography by several researchers (4,7-14). In 1979, we reported on a new surface wave launching device called a surfatron (15), which could be used to generate plasmas in different gases at any pressure up to atmospheric pressure. This device has been used as an excitation source for a gas chromatography detector (16-22), for a supercritical fluid chromatography detector (23-25) and for the analysis of liquid samples (26-29).
This paper presents work performed in our laboratory with a surface wave plasma as an element selective chromatographic detector. In the first part, we will briefly describe the essential features of a surface wave plasma, we will recall the emission features of the helium plasma as an excitation source for non-metals analysis, and will then describe the effect of several operating parameters (flow rate, power and

[1]Current address: CANMET, 555 Booth Street, Ottawa, Ontario, Canada K1A OG1

discharge tube geometry) on the emission intensity of non-metals. The related changes in electron density, excitation and gas temperature will also be investigated. Finally, we will present a few applications of this detector in gas chromatography.

Experimental Section

The instrumental arrangement (Figure 1) consists essentially of three parts: the plasma generation unit, the sample introduction systems and the detection system.

Plasma Generation Unit. The previously described surfatron (*17*) was used to generate the helium plasma. Two models were used in this work, each of them designed to match the frequency of the microwave generator. For power levels lower than 200 W, a 2450 MHz MK-III Microtron power generator (Electro-Medical Supplies, London, England) was used, while a 915 MHz custom-made Cober generator (Cober, Stamford, Conn.) provided power from 150 to 500 W. The power for each generator was calibrated by using a directional coupler line (Narda, Hauppage, NY, Model 3020A was used at 915 MHz, and model 3022 at 2450 MHz) and a thermistor power meter (Hewlett-Packard, Model HP432a) (*30*). The power was transfered from the generator to the surfatron by a 0.5 m length coaxial cable (Type RG-214/U).

The discharge tubes were quartz capillary tubes ranging from 6 mm o.d. / 1 mm i.d. to 8 mm o.d. / 6 mm i.d.. Both surfatrons could accept any of these tubes.

Sample Introduction Systems. For the optimization work, the sample introduction system depicted in Figure 2 was used. The helium flow was split so that the "internal" gas line (I) carries flow rates from 15 to 300 mL/min. The sample line (S) flow rate was kept constant at 1 mL/min to insure a constant introduction rate of vapor coming from the sample tube (T) (Figure 2a). When a two tube discharge torch configuration was used, a third "external" gas line (E) carried flow rates from 0 to 300 mL/min (Figure 2b). The sample tube is a capillary evaporation tube of pyrex (i.d. 1.5 mm, o.d. 6 mm) filled with an appropriate organic solvent.

The chromatographic work was performed with the following apparatus. A gas chromatograph (Hewlett-Packard, Model 5750) was modified and the microwave induced plasma was adapted as described in reference (*17*) . The column was a glass tube (length 235 cm, i.d. 2 mm) packed with 5% SE-30 coated Chromosorb G, AW, DMCS (80-100 mesh). The helium gas flow rate was 50 mL/min. This chromatograph was used for the analysis of some pesticides. A capillary gas chromatograph (Perkin-Elmer, Model Sigma 3) was also used for the chlorinated pesticide analyses. The column was a SE-30 fused silica capillary column (length 25 m, i.d. 0.25 mm). The carrier gas flow rate was 1 mL/min and a split ratio of 1/50 was used at the injector. At the outlet of the column a make-up flow rate of 50 mL/min was added to the carrier gas.

The chemicals used were supplied by Aldrich (Milwaukee, Wisconsin, USA) and Chem Service (Westchester, PA, USA) and were used without further purification.

Detection System. The plasma was viewed axially and the emitted light was focused on the entrance slit of the monochromator by a 10 cm focal length quartz lens to produce a 1:1 image. For the optimization work, the monochromator was a 1 m focal length Jarrell-Ash, model 75-150, with a 1180 grooves / mm grating blazed at 250 nm. The entrance and exit slits were kept constant at 0.020 mm. For the diagnostic studies and the chromatographic work a Jobin-Yvon monochromator, model HR1000, with a 1 m focal length and a 2400 grooves / mm grating optimized in the UV, was used. The entrance and exit slits were kept constant at 0.010 mm for the diagnostic studies and set at 0.040 mm for the chromatographic work. The

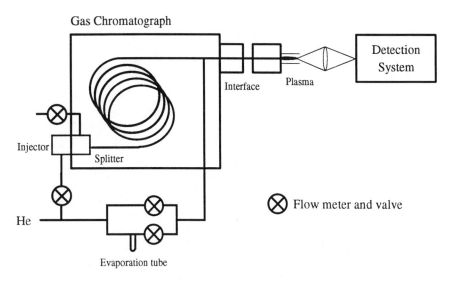

Figure 1. Schematic diagram of the instrumental set-up.

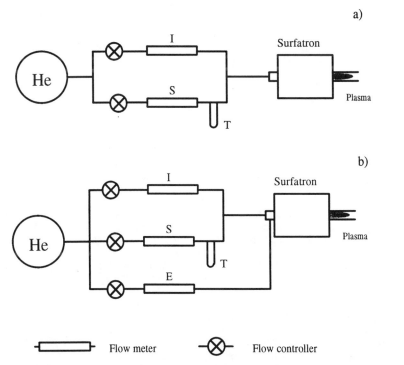

Figure 2. Schematic diagram of the gas control system. He: Helium, I: Internal or plasma gas line, S: Sample carrier gas line,T: Sample evaporation tube, E: External plasma gas line.

photomultiplier was a Hamamatsu R-636 with a Hewlett-Packard 5551A high voltage power supply. The photomultiplier current was monitored by a Keithley model S414 picoammeter and a Brinkman model 2541 recorder was used to record the signal.

Surface Wave Induced Plasmas

There are several ways by which microwave energy can be transferred to sustain a plasma. In general the electromagnetic energy is transmitted to the plasma via the microwave electric field. The electric field accelerates the electrons that dissipate the microwave energy through elastic and inelastic collisions with neutral and ionic species. The electric field can be applied to a plasma within an electromagnetic resonant cavity (6) or it can come from a wave travelling along the plasma column. If the travelling wave uses the plasma column as the sole propagating medium, the plasma produced is called a surface wave induced plasma or surface wave plasma (SWP). Surface wave plasmas have a number of unique properties when compared to resonant cavity produced plasmas: i) single mode propagation can be achieved and the resultant plasmas are very stable and reproducible, ii) for a given mode the wave propagation is independent of the configuration of the launcher and iii) the frequency bandwidth of the launching structure is large. The properties of the surface wave plasmas have been extensively studied and their modelling is well developed for plasmas operated under reduced pressure (31). The number of surface wave induced helium plasmas studied at atmospheric pressure is limited and the modelling of these plasmas is still in its infancy.

Helium Surface Wave Plasma Characteristics

The helium surface wave plasma at atmospheric pressure is typically operated in a narrow discharge tube (2-4 mm i.d.) and the absorbed power is less than 200 W (16-22, 26-29, 32). Only a limited amount of data about the physical characteristics of the helium plasma are available (32). A helium plasma operated at atmospheric pressure in a 2 mm i.d. discharge tube, at a flow rate of 0.5 L/min and at 82 W of absorbed power was characterized in terms of excitation temperature, rotational temperatures (N_2 and N_2^+) and electron density at various frequencies. The excitation temperature was about 3000 K, whereas the N_2 and N_2^+ rotational temperatures were 2000 K. The electron density was about 1 x 10^{14} e$^-$/cm^3. The effect of frequency is weak, if not absent, under these conditions and for the frequency range investigated (32).

   This work has now been extended to helium plasmas operated at 915 and 2450 MHz at higher power levels (up to 300 W), at lower flow rates (< 200 mL/min) and in various discharge tubes (i.d. 3-6 mm, two section tubes). No frequency effect was observed between 915 and 2450 MHz.

   Calculations based on Boltzman diagrams and the two-line method were made to estimate the excitation temperature. The temperature was estimated from helium and iron emission lines. The excitation temperature derived from the helium line intensities was 3900 K, while the iron line intensities gave a temperature of 3600 K. Both helium and iron probes showed no significant variations of the excitation temperatures over the range of power, flow rate and discharge tube size and geometry under investigation.

   The rotational temperatures were measured using N_2 and N_2^+ as probes. The temperature remained constant when the flow rate, the discharge tube diameter or geometry were changed. However an increase in rotational temperature from 1300 to 2000 K was observed when the power was increased from 40 to 300 W. A better atomization at higher power levels can therefore be expected.

   The electron density was evaluated from the Stark broadening of the H$_\beta$ line. There is no significant change of the electron density with the change in flow rate and in absorbed power. However, a significant increase was observed when shifting from the 6 mm i.d. discharge tube to the 3 mm i .d. tube (diameters taken in the plasma zone). The plasma density increased by about 37% and a value of 1.1×10^{14} e⁻/cm³ was obtained for the smallest discharge tube.

Emission Spectra of Non-Metals in the Helium Surface Wave Plasma

The emission spectral features of non-metals introduced into a helium surface wave induced plasma at atmospheric pressure have been described (*33*). For carbon, phosphorus and fluorine only atomic lines were observed. All the other elements investigated show the emission of several ion and atom lines. The general emission characteristics of the helium surface wave induced plasma were similar to the emission profiles obtained with a plasma sustained in a Beenakker type cavity.

Effect of the Plasma Operating Parameters on the Emission Intensity

The effect of the gas flow rate, the discharge tube geometry and the absorbed microwave power on the line emission of non-metals, the background intensity and the noise on the background intensity was studied over a large variety of operating conditions. The elements under study and the wavelengths which were used are reported in Table I.

Helium Flow Rate and Discharge Tube Geometry. One of the common well documented features of the various MIPs is the difficulty of introducing samples into the plasma to obtain efficient atomization and excitation. The small diameter of the MIPs (1-5 mm) as compared to the ICP makes the sample penetration difficult. In order to improve the atomization and excitation efficiency various discharge tube configurations were used. A series of common straight discharge tubes with internal diameters ranging from 1 to 8 mm were first used. In general the line intensity of the most sensitive lines for the non-metals decreases with an increase of the flow rate as previously reported in reference (*17*).

A two concentric tube discharge torch (Figure 3a) similar to a two tube mini-ICP torch was also used. In this arrangement, an internal gas flow (I) brings the sample to the plasma through an internal alumina tube (i.d. 1 mm, o.d. 2 mm) while the external gas flow (E) was tangentially injected between the internal and the external quartz tubes. The external tube had a 3 mm i.d. Typical results of the change in net intensity emitted by the chlorine ion line at 479.45 nm as a function of the flow rate changes are shown in Figure 4 at an absorbed power of 60 W. When the internal gas flow rate was about 60 mL/min an increase of the line intensity was observed when the external flow was increased up to a flow rate of 60 mL/min. At high flow rates, a decrease of the line intensity was observed. With a lower internal gas flow rate of 25 mL/min, the same trend was obtained, with a shift of the maximum to a higher external gas flow rate of 170 mL/min. A further decrease of the internal gas flow rate to about 10 mL/min shows that the overall line intensity is lower in intensity and that a maximum in emission was obtained for a flow rate of 180 mL/min. At this power level, it was not possible to sustain a stable plasma at higher flow rates. By changing the relative position of the plasma to the internal injection tube from 1 to 8 cm no significant change in intensity was observed.

This two tube torch arrangement did not prove to be more efficient in terms of sample introduction whatever the flow rate conditions. On the contrary, in terms of emitted intensity, the best emission conditions were found with no external gas flow rate. Also the observed intensity was higher than that obtained with a simple straight tube (Figure 5). A change in diameter between the sample introduction tube and the plasma discharge tube seems beneficial to the emission intensity. This assertion was

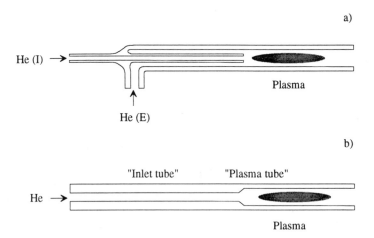

Figure 3. Discharge tube torches. a) ICP type mini-torch, b) "Two section tube"torch.

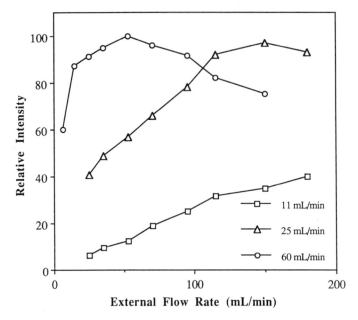

Figure 4. Effect of the external and the internal gas flow rates on the chlorine II emission line intensity. External quartz tube: 6 mm o.d., 3 mm i.d.. Internal alumina tube: 2 mm o.d., 1 mm i.d.. Absorbed power: 60 W. Internal flow rate: 11 mL/min, 25 mL/min, 60 mL/min.

confirmed by working with a "two section" discharge tube (Figure 3b) in which the sample and the plasma gas come from a narrow section ("inlet tube") and the plasma is located in a larger section ("plasma tube") . The narrow section ("inlet tube") of the tube was varied from 1 to 3 mm i.d. and the larger section ("plasma tube") was varied from 3 to 6 mm i.d..

Figure 6 shows the flow rate dependance of the chlorine signal observed when a "two section" discharge tube was used. Each curve shows a maximum whose position is shifted to higher flow rates when the difference in internal diameter between the "inlet tube" and the "plasma tube" becomes smaller. The decrease in intensity observed at higher flow rates is probably due to a lower residence time and a dilution effect as the flow rate increases. At lower flow rates, the lower intensity probably results from a less efficient sample penetration into the helium plasma.

Figure 7 shows how the intensity of the line is influenced by the change of the internal diameter of the "plasma tube" and the relative size of the "inlet tube" diameter to the "plasma tube" diameter . For the two plasma diameters, the "two section tubes" provide a more efficient excitation of the analyte which confirms the tendencies noted above. However, the signal is the largest for the smallest "plasma tube". The background noise remained constant. As mentioned earlier, no significant differences were found between the excitation temperature, the rotational temperature and the electron density in plasmas generated in a straight or a "two section" discharge tube for which the "plasma tube" internal diameter was the same (3 mm i.d.). The observed change in intensities can therefore only be explained by a more efficient penetration of the sample into the plasma in the case of the "two section tube".

Comparison between Atomic and Ionic Lines. The experiments performed with iodine, sulfur, bromine and carbon showed the same trends as the chlorine in relation to the plasma discharge tube geometry, whatever was the power level. A "two section" plasma tube with an 1 mm i.d. for the "inlet tube" and a 3 mm i.d. for the "plasma tube" was used. With this tube the optimum operating flow rate varies with the monitored species. Figure 8 shows that the intensity for all atomic lines exhibits approximatively the same behavior, showing a maximum around 20-25 mL/min. It is interesting to notice that the flow rate dependency of the emission intensity is similar for carbon and iodine, and for these elements a slower decrease in intensity is observed when the flow rate is increased. However, the sulfur and chlorine lines decrease in intensity more rapidly. This seems to be in agreement with the difference in excitation energy for the different transitions (Table I).

The emission behavior of the ion lines changes more drastically from one element to another (Figure 9). The position of the curve maximum suggests that chlorine is more efficiently excited at higher flow rates than iodine or sulfur. From the data in Table I, one can see that only the difference in the first ionization potential is in agreement with the observed data. Chlorine is the most difficult element to ionize and is efficiently excited at higher flow rates for which a more effective mixing of the sample and the plasma occurs. This information also suggests that a two-step excitation mechanism (ionization $X \rightarrow X^+$ and excitation $X^+ \rightarrow X^{+*}$) is certainly involved with ionization being the most critical step.

From the data in Figure 8 and Figure 9, we can conclude that one must choose carefully the flow rate conditions for the element of interest. In case of multi-element analyses, the flow rate must be adjusted at a compromise value for the different lines especially when dealing with ionic lines, whereas with the atomic lines the compromise conditions are less critical.

Effect of Absorbed Power. With the surfatron, the experiments performed show that the intensity increases with an increase in absorbed power. Moreover, the signal-to-background noise ratio which is a better indication of the analytical

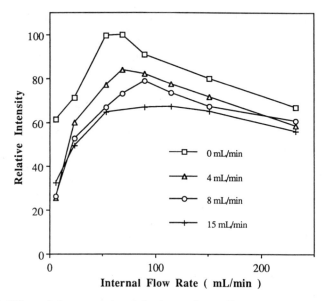

Figure 5. Effect of the external and the internal gas flow rates on the chlorine II emission line intensity. External quartz tube: 6 mm o.d., 3 mm i.d.. Internal alumina tube: 2 mm o.d., 1 mm i.d.. Absorbed power: 60 W. External gas flow rate: 0 mL/min, 4 mL/min, 8 mL/min and 15 mL/min.

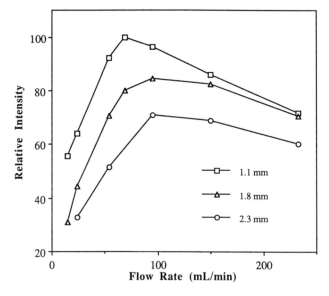

Figure 6. Effect of the "inlet tube" internal diameter on the chlorine II emission line intensity. "Plasma quartz tube": 6 mm o.d., 3 mm i.d.. "Inlet quartz tube": 6 mm o.d.. Absorbed power: 60 W. "Inlet tube" i.d.: 1.1, 1.8 and 2.3 mm.

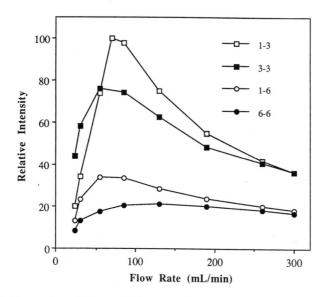

Figure 7. Comparison between different straight discharge tubes and two-section tubes on the chlorine II line intensity. Absorbed power: 150 W. Straight tubes: 3-3 = 6 mm o.d., 3 mm i.d., 6-6 = 8 mm o.d., 6 mm i.d.. Two section tubes: 1-3 = 6 mm o.d., "inlet tube": 1 mm i.d. and "plasma tube": 3 mm i.d., 1-6 = 8 mm o.d., "inlet tube": 1 mm i.d. and "plasma tube": 6 mm i.d..

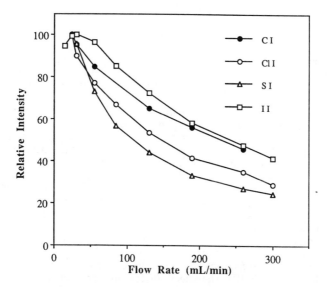

Figure 8. Flow rate dependance of the atom line intensity. Two section tube: "inlet tube": 1 mm i.d., "plasma tube": 3 mm i.d., Absorbed Power: 60 W. C I: 247.86 nm, Cl I: 725.67 nm, S I: 469.55 nm and I I: 206.16 nm.

Table I. First Ionization Potentials and Excitation Energies for Non-metals

	Wavelength studied (nm)	First Ionization Potential (eV)	Excitation Energy (eV)	Total Energy[a] involved (eV)
Carbon I	247.86	11.26	7.68	
Bromine II	470.49	11.81	14.28	26.09
Bromine II	478.55		14.23	26.04
Chlorine I	725.67	12.97	10.62	
Chlorine II	479.45		15.89	28.86
Iodine I	206.16	10.45	6.95	
Iodine II	516.12		12.45	22.90
Phosphorus I	253.56	10.49	7.18	
Sulfur I	469.55	10.36	9.16	
Sulfur II	545.39		15.88	26.24

[a] Total energy = First Ionization Potential + Excitation Energy of the Ion Line

performance in terms of detection limits, also increases from 100 to 300 W by a factor of 10 to 25, depending on the element and line considered (Figure 10). This means that the detection limit is likely to be improved in going from the original 50-100 W range (used in most work on non-metal detection by MIP atomic emission) up to 300 W. For example, the detection limit for chlorine decreases from 20 pg/s with an absorbed power of 100 W to 4 pg/s at a power level of 230 W. The main restriction to a further increase in power is the quartz discharge tube lifetime.

Applications in Gas Chromatography

Response Factor and Interhalogen Selectivity. The most sensitive and selective halogen detector presently available for gas chromatography is the electron capture detector (ECD). In order to assess the usefulness of the surface wave plasma atomic emission detector for gas chromatography, the relative response factors for several chlorobenzene compounds were determined with both detectors using similar analytical conditions. The values obtained for the response factors are shown in Table II. The response factors were normalized relative to 1,2,4-trichlorobenzene.

The maximum change obtained in response factors with the surface wave plasma atomic emission detector was 9%, with most values for the various compounds being within 5% of the reference compound. However, very large differences in the response factors were obtained for the same compounds with the ECD. Differences as large as 10^4 are observed between chlorobenzene and hexachlorobenzene. The values obtained with the ECD are typical for this type of detector as reported in reference *34* and are related to the difference in inductive effects of chlorine and in electron density changes of the various molecules. The small change in response factors with molecular structure is a definite advantage for the surface wave plasma atomic emission detector in gas chromatography. The same type of results were also observed for helium plasmas generated by different means.

The selectivities for both detectors were evaluated for various halogen containing aromatic molecules (Table III). The selectivity for chlorine over carbon is very high for the ECD. The selectivity for the surface wave plasma atomic emission detector was obtained with a monochromator of medium resolution and can be improved by

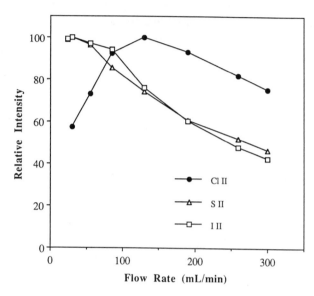

Figure 9. Flow rate dependance of the ion line intensity. Two section tube: "inlet tube": 1 mm i.d., "plasma tube": 3 mm i.d., Absorbed Power: 60 W. C II: 479.45 nm, S II: 545.39 nm and I II: 516.12 nm.

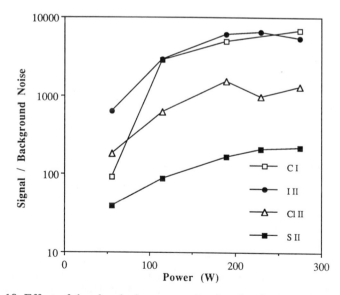

Figure 10. Effect of the absorbed power on the signal to background noise ratio. Two section tube: "inlet tube": 1 mm i.d., "plasma tube": 3 mm i.d., Plasma gas flow rate: 70 mL/min.

Table II. Relative Response Factors[a] of the Microwave Emission Detector and the Electron Capture Detector for Chlorobenzenes

Compound	Microwave[b] detector	Electron capture detector
Chlorobenzene	1.05	8.26×10^{-4}
1,2-Dichlorobenzene	1.07	1.62×10^{-1}
1,3-Dichlorobenzene	1.09	1.83×10^{-1}
1,4-Dichlorobenzene	0.99	7.82×10^{-2}
1,2,4-Trichlorobenzene	1.00	1.00
1,3,5-Trichlorobenzene	0.99	1.28
1,2,3,4-Trichlorobenzene	1.05	3.93
1,2,3,5-Trichlorobenzene	1.01	2.60
1,2,4,5-Tetrachlorobenzene	0.98	1.42
Pentachlorobenzene	1.02	11.0
Hexachlorobenzene	0.95	14.1

[a]The response factors were calculated relative to 1,2,4-trichlorobenzene as a standard.
[b]Detection based on the emission of the chlorine ion line at 479.45 nm.

Table III. Selectivity[a] of the Electron Capture Detector and the Microwave Emission Detector

	Cl / C	Cl / Br	Cl / I
n-Hexane	1.2×10^6		
Benzene	8.8×10^5		
Dibromomethane		4.5×10^{-2}	
Bromomethane		8.1×10^{-1}	
Bromobenzene		1.3×10^{-1}	
Iodoethane			2.0×10^{-1}
Iodopropane			1.5×10^{-1}
Iodobenzene			5.1×10^{-2}
Microwave detector [b]	600	1100	800

[a] The selectivity was calculated using the response of 1,2,4-trichlorobenzene as a standard.
[b] From reference (17).

increasing the resolution or by using a real time background correction by wavelength modulation (35).

If we compare the selectivities of both detectors for chlorine over bromine and chlorine over iodine, we observe only slight variations for the surface wave atomic emission detector, whereas for the ECD the variations are considerable. The interhalogen selectivity of the surface wave plasma atomic emission detector is certainly one of the important features of this detector.

Analysis of Pesticides. A number of organochlorine and organophosphorus pesticides were analyzed on a fused silica capillary column by gas chromatography. The detection limits (2 σ) were measured and are reported in Table IV. The detection limits obtained are equivalent to those reported by Rivière et al.(*20*). In some cases, the use of a surface wave plasma operated under reduced pressure provided improved detection limits. However, the improvement is marginal and does not justify the use of a cumbersome vacuum system.

Table IV. Detection limits for pesticide analysis

Pesticide	Carbon (pg/s)	Chlorine (pg/s)	Phosphorus (pg/s)	Sulfur (pg/s)
Lindane	7	10		
Heptachlor	13	13		
Aldrin	17	30		
DDE	27	20		
Dieldrin	17	20		
DDT	13	13		
Phorate	9		3	33
Diazinon	9		3	20
Ronnel	15		1	27
Cygon	14		3	40

Wavelength: C I: 247.86 nm, Cl II 479.45 nm, P I 253.56 nm, S II 545.39 nm

Two pesticides were analyzed in environmental samples. Bromacil (Figure 11a) was extracted from soil samples with ethylacetate and a preconcentration factor of 50 was used. The analysis was performed on a packed column with a NPD detector (nitrogen detection) and the surface wave plasma atomic emission detector. The bromine II line at 478.55 nm was used and an example of a chromatogram is shown in Figure 12. Four different samples were analyzed and the results obtained for the analyzed extracts indicate an excellent agreement between the two detectors. The precision with the surface wave plasma atomic emission detector was 2% and the differences in concentration between the two series of measurements were less than 5%.

Picloram (Figure 11b) was analyzed in river water samples. A fraction of 800 mL of water was extracted with ether. After extraction, picloram was converted into its methyl ester in the presence of boron trifluoride. The ester was extracted with hexane and the resulting solution was analyzed by gas chromatography. A total preconcentration factor of 800 was used. The analysis was performed with an ECD detector and with the surface wave induced plasma atomic emission detector. The chlorine II line at 479.45 nm was used and an example of a chromatogram is shown in Figure 13. The results obtained (concentration 0.3 ppm) are in good agreement (less than 2% difference) with the ECD data.

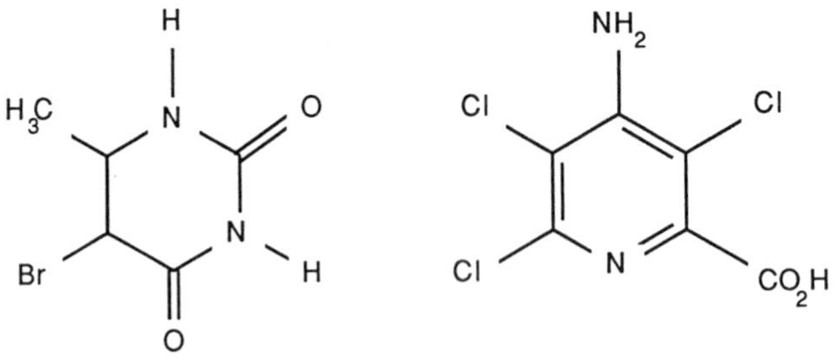

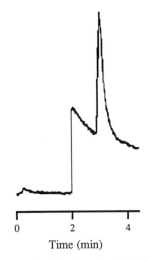

a)   b)

Figure 11. Pesticide formula. a) Bromacil and b) Picloram.

Time (min)

Figure 12. Analysis of Bromacil. Bromine II: 478.55 nm, Two section tube: "inlet tube": 1 mm i.d., "plasma tube": 3 mm i.d., Absorbed power: 230 W. Gas flow rate: 50 mL/min. Chromatographic conditions: Packed column: 1.8 m, 2 mm i.d., SE-30 5% on Chromosorb W-HP 80/100 mesh. Isothermal separation. Temperature: Oven: 220 °C, Injector: 280 °C and Interface: 350 °C.

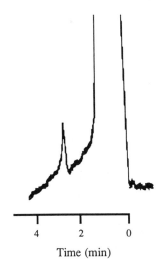

4   2   0

Time (min)

Figure 13. Analysis of Picloram. Chlorine II: 479.45 nm, Two section tube: "inlet tube": 1 mm i.d., "plasma tube": 3 mm i.d., Absorbed power: 230 W. Gas flow rate: 50 mL/min. Chromatographic conditions: Packed column: 1.8 m, 2 mm i.d., SE-30 5% on Chromosorb W-HP 80/100 mesh. Isothermal separation. Temperature: Oven: 200 °C, Injector: 270 °C and Interface: 350 °C.

Acknowledgements

We wish to thank the Natural Sciences and Engineering Research Council of Canada (NSERCC) and le Fonds pour la Formation de Chercheurs et Aide à la Recherche du Québec (FCAR) for the financial support of this work.

Literature Cited

1) McCormack, A.J.; Tong, S.C. and Cooke, W.D, Anal. Chem., **1965**, *37*, 1470.
2) Bache, C.A. and Lisk, D.J., Anal. Chem., **1967**, *39*, 786.
3) McLean, W.R.; Stanton, D.L. and Penketh, G.E., Analyst, **1973**, *98*, 432.
4) Van Dalen, J.P.J. ; De Lezenne Coulander, P.A. and de Galan, L., Anal. Chim. Acta, **1977**, *94*, 1.
5) Zander, A.T. and Hieftje, G.M., Appl. Spectros., **1981**, *35*, 357.
6) Beenakker, C.I.M., Spectrochim. Acta, **1977**, *32B*, 173.
7) Quimby, B.D.; Uden, P.C. and Barnes, R.M., Anal. Chem., **1978**, *50*, 2112.
8) Quimby, B.D.; Delaney, M.F.; Uden, P.C. and Barnes, R.M., Anal. Chem., **1979**, *51*, 875.
9) Tanabe, K.; Haraguchi, H. and Fuwa, K., Spectrochim. Acta, **1981**, *36B*, 633.
10) Estes, S.A. ; Uden, P.C. and Barnes, R.M., Anal. Chem., **1981**, *53*, 1829.
11) Risby, T. H. and Talmi, Y., CRC Crit. Rev. Anal. Chem., **1983**, *14*, 231.
12) Matousek, J.P. ; Orr, B.J. and Selby, M., Prog. Anal. Atom. Spectros., **1984**, *7*, 275.
13) Deruaz, D. and Mermet, J.M., Analusis, **1986**, *14*, 107.
14) Uden, P.C. ; Yoo, Y. ; Wang, T. and Cheng, Z., J. Chrom., **1989**, *468*, 319.
15) Hubert, J.; Moisan, M. and Ricard, A., Spectrochim. Acta, **1979**, *34B*, 1.

16) Hanai, T. ; Coulombe, S. ; M. Moisan and J. Hubert, In *Developments in Atomic Plasma Spectrochemical Analysis*, Barnes, R.M., Heyden, London, **1981**, pp. 337-344.

17) Chevrier, G. ; Hanai, T. ; Tran, C.K. and Hubert, J., Can. J. Chem., **1982**, *60*, 898.

18) Takigawa, Y. ; Hanai, T. and Hubert, J., J. High Resol. Chrom., **1986**, *9*, 698.

19) Hanai, T. and Hubert, J., In *Proceedings of the 8th International Symposium on Plasma Chemistry*, Akashi, A.and Kingara, A., IUPAC, Tokyo, Japan, **1987**, 8, pp. 383-384.

20) Rivière, B. ; Mermet, J.M. and Deruaz, D., J. Anal. Atom. Spectrom., **1987**, *2*, 705.

21) Besner, A. and Hubert, J., J. Anal. Atom. Spectrom., **1988**, *3*, 381.

22) Hubert, J. ; Lauzon, L. and Tran K.C., J. Anal. Atom. Spectros., **1988**, *3*, 901.

23) Galante, L.J. ; Selby, M. ; Luffer, D.R. ; Hieftje, G.M. and Novotny, M. and, Anal. Chem., **1988**, *60*, 1370.

24) Luffer, D.R. ; Galante, L.J. ; David, P.A. ; Novotny, M. and Hieftje, G.M., Anal. Chem., **1988**, *60*, 1365.

25) Rivière, B. ; Mermet, J.M. and Deruaz, D., J. Anal. Atom. Spectrom., **1988**, *3*, 551.

26) Abdallah, M.H. ; Coulombe, S. ; Mermet, J.M. and Hubert, J., Spectrochim. Acta, **1982**, *37B*, 583.

27) Selby, M. ; Rezaaiyann, R. and Hieftje, G., Appl. Spectros., **1987**, *41*, 749.

28) Selby, M. ; Rezaaiyann, R. and Hieftje, G., Appl. Spectros., **1987**, *41*, 761.

29) Galante, L.J. ; Selby, M. and Hieftje, G.M., Applied Spectros., **1988**, *4*, 559.

30) Hubert,J. ; Moisan, M. and Zakrzewski, Z., Spectrochim. Acta, **1986**, *41B*, 205.

31) Moisan, M. ; Ferreira, C.M. ; Hajlaoui, Y. ; Henry, D. ; Hubert, J. ; Pantel, R. ; Ricard, A. and Zakrzewski, Z., Revue Phys. Appl., **1982**, *17*, 707.

32) Besner, A. ; Moisan, M. and Hubert, J., J. Anal. Atom. Spectros., **1988**, *3*, 863.

33) Hubert, J. ; Tra, H. ; Tran, K.C. and Baudais, F., Applied Spectros., **1986**, *40*, 759.

34) Zlatkis, A. and Poole, C.F., In *Electron Capture, Theory and Practice in Chromatography*, Elsevier Sci. Pub. Com., Amsterdam, Oxford, New-York, **1981** pp. 208-261.

35) Houpt, P.M., Anal. Chim. Acta, **1976**, *86*, 129.

RECEIVED May 6, 1991

Chapter 12

Helium Discharge Detector for Gas Chromatography

Gary W. Rice

Department of Chemistry, College of William and Mary,
Williamsburg, VA 23185

An overview of the design, analytical figures of merit,
recent applications, and limitations of a helium discharge
detector (HDD) for element selective detection in gas
chromatography is presented. The operational features,
spectral characteristics, excitation temperatures, and
considerations for the means of analyte excitation relative
to the microwave induced plasma are discussed.
Utilization of alternative analytical wavelengths in the
near IR is evaluated for improved selectivity of several
elements traditionally detected in the UV. Potential
improvements in the HDD are considered.

The innumerable compounds that can potentially be detected in a
chromatographic analysis has led to an increasing need for element
specific, multi-element detectors for gas chromatography (GC). This
interest has magnified over the past thirty years to the extent that
commercial instrumentation is now readily available. The stringent
criteria under which these detectors are evaluated include: 1) limits of
detection which should equal or surpass those of a flame ionization
detector (~10 pg); 2) wide linear range of response; 3) tolerance to the
passage of eluting solvent; 4) element selectivity over non-analyte
species; 5) predictable response which is independent of the compound
type; and 6) simple operation and maintenance.

The helium microwave induced plasma (MIP) has clearly developed
into a versatile, element-selective detector which for the most part
meets the afore mentioned criteria. The Hewlett-Packard GC-Atomic
Emission Detector (GC-AED) allows several elements from eluents in a
GC chromatogram to be monitored simultaneously using a movable
photodiode array detector (1). This mode of detection also allows for

0097–6156/92/0479–0205$06.00/0
© 1992 American Chemical Society

rapid background correction and elemental verification through three dimensional time-wavelength-intensity profiles *(2)*. The GC-AED system with continued refinements will undoubtedly become the detector of choice for laboratories where complex and numerous determinations are required for a variety of elemental compositions.

We have continued developing and evaluating a helium discharge detector (HDD) for GC which possesses many of the attributes of the MIP but whose construction, mode of operation, and means of analyte excitation are noticeably different. The HDD is in a stage of infancy relative to the MIP, having been designed, developed, and evaluated only over the past seven years by a few laboratories. A system similar in design to that used in this laboratory is currently marketed by CETAC Technologies, and modifications for other applications have recently been proposed and described *(3,4, Diehl,J. University of North Dakota, personal communication, 1989)*.

This chapter describes the design, operation, and a short overview of past observations which validate the HDD as a versatile, element selective detector for GC. Recent findings on the capabilities of the HDD for detection of analyte species in the near IR region of the spectrum are also discussed, as well as plausible explanations for the mode of analyte excitation.

Instrumentation

A schematic diagram of the HDD is shown in Figure 1. The configuration is very similar to the original design *(5)*. The electrodeless type discharge is coupled through a 3 mm o.d. by 1 mm i.d. quartz tube from a 2 cm length cylindrical stainless steel electrode connected to a high voltage, variable frequency power supply. The tuned frequency at which minimum power reflectance occurs is 176 KHz by comparison with 26 KHz in the original system. This change is primarily due to the use of a commercially available high voltage, high frequency power supply, which couples the energy applied to the helium discharge more efficiently at higher frequencies. The region of the quartz tube surrounded by the electrode is designated as the primary discharge region. The 3 cm extension of the quartz tube above the electrode is capacitively coupled to a grounding electrode. This region is designated as the secondary discharge region, from which all optical emission is monitored. The tube is mounted and sealed on a thermally regulated, stainless steel heater block to maintain the GC capillary column at or above the temperature of the GC oven. The entire electrode assembly is surrounded by ceramic insulators primarily as a safety precaution. The GC capillary column is inserted and sealed at the base of the heater block assembly and positioned so that the exit is located at the beginning of the secondary discharge region. The polymeric coating on the last 8-10 cm of the column is removed to prevent deposits from occurring in the discharge tube from interactions with the primary discharge. The helium flow used to sustain the discharge enters through a port on the side of the heater block. The entire assembly is mounted directly on the top of the GC oven.

A schematic diagram of the entire GC-HDD system is shown in Figure 2. The flow of the chromatographic grade helium (99.9999%, Air Products) is regulated and monitored with a digital flow meter and then passed through a heated catalytic purifier (Supelco) to remove trace impurities, e.g., O_2, H_2O, prior to entering the discharge assembly. Optical emission from the secondary discharge is focussed through a 50 mm focal length CaF_2 lens into a 0.5 m monochromator. The desired wavelengths of emission are detected by a photomultiplier tube and monitored via conventional spectrophotometric instrumentation. Optical filters placed between the discharge and focussing lens may be used to eliminate second and/or third order spectral interferences when monitoring the visible and near IR spectral regions. Amplified signals can be simultaneously recorded on a strip chart recorder or stored on an interfaced IBM PS/2 PC. Typical operating parameters and instrument specifications are summarized in Table I.

Table I. Instrument Specifications and Operating Parameters for the GC-HDD System.

Discharge System		
Incident Power	60 Watts	
Voltage	6500 RMS	ENI, Inc.
Frequency	176 KHz	Model HPG-2
He Flow Rate	60 mL/min	
Spectrometric System		
Monochromator	0.5 m (Minuteman)	
Bandpass/slitwidths	0.35 nm/ 150 µm	
PMT Voltage	1060 V (Hamamatsu)	
Amplifier	Keithley Model 485	
Signal Processing	IBM PS/2 Model 30 E-21 with Axxiom Data System	
Chromatographic System		
GC	Carlo Erba 4700 with on-column injection	
Carrier Gas	Helium	
Column (typical)	DB-5, 30 m (J&W Scientific)	

Fundamental Characteristics of the HDD

Operation. There are several features of the helium discharge which are unique relative to conventional plasma systems. The visual

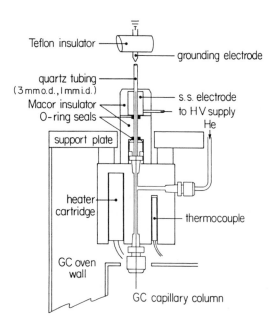

Figure 1. Schematic diagram of the helium discharge detector.

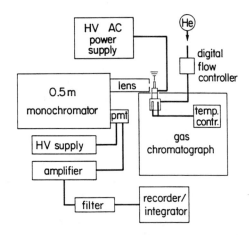

Figure 2. Schematic diagram of the GC-HDD system.

intensity of the salmon colored discharge is less than an ordinary candle. The radiative temperature generated from the quartz walls in the secondary discharge region is relatively low (300-350 °C). Thus, protective measures against radiative hazards and cooling of the system have not been required. The discharge is self-initiating and requires no external ignition, e.g.,Tesla coil. The efficiency of power coupling for producing the helium discharge is highly dependent on the operational frequency (5% variations in the frequency can reduce the load power as much as 30%). The end of the quartz tube can be sealed and the entire gas train pressurized with helium when not in use. This precaution allows for a constant background signal to be attained within minutes of ignition since no atmospheric contaminants have to be purged.

The most distinctive feature of the HDD is that the GC eluents are introduced into the secondary discharge region instead of directly into the plasma source as is the case with conventional MIP systems. The secondary discharge is noticeably quenched by excessive sample or solvent loading (>0.01 μL volumes); however, total extinguishment under normal chromatographic conditions is not possible. The exit of the GC capillary column is at the end of the primary discharge. Thus the primary discharge, which is unperturbed by the chromatographic process, continually reestablishes the secondary discharge. Excessive volumes of solvent (3-5 μL) may result in carbonization on the quartz walls as the rate of carbon formation exceeds the rate of removal necessary to prevent deposition. A stable baseline emission signal is quickly reestablished after the elution of typical solvent volumes in chromatographic processes (0.1-2 μL). The obvious advantage is that solvent venting hardware typically needed in MIP systems to prevent total shutdown of the detector is not required in the HDD system. No degradation in the chromatographic peak shapes has been observed when compared to the performance of a conventional FID detector on the same GC system.

Spectral Characteristics.   The most prominent background emissions which have been characterized in the secondary discharge region of the HDD include molecular band emission from CO, NH, OH, and N_2 in the UV/VIS regions of the spectrum. These background emissions are typically observed in MIP systems as well. The intensity of these emissions will depend on the level of trace impurities in the helium support gas. Other emissions, such as atomic emission from neon, can be observed depending on the source of helium. Emission from species such as N_2 and CO is easily rationalized by the fact that the energy required to produce the observed molecular band systems coincides with the 2^1S and 2^3S He metastable states, thus allowing for efficient resonance energy transfer processes to occur.

The variety of emissions in the UV/VIS region of the background spectrum has a significant impact on the selective detection of specific atomic emissions for a given element. On the other hand, background emission above 500 nm is for the most part restricted to atomic emissions from He as well as H, N, and O from impurity gases. The impact of the background emissions on the selection of analytical wavelengths for good element selectivity is considered in the analytical performance section of this chapter.

Excitation Temperature. The excitation temperatures associated with the HDD have recently been determined by measuring the relative intensities of characteristic He transitions present in the background spectrum *(6)*. Each transition will follow a Boltzmann distribution if one assumes that the discharge is in a state of local thermodynamic equilibrium with the inner walls of the quartz tube. Temperatures can then be determined from the slopes of the Boltzmann plots. The intensities of eight He transitions were measured at various power levels, flow rates, and observation heights within the secondary discharge region. All intensities were corrected for grating and detector efficiencies at each respective wavelength.

An excitation temperature of approximately 2500 K was determined under normal operating conditions (60 W power, 60 mL/min flow rate). Little variation was observed relative to cross sections of measured points from the bottom to the top of the secondary discharge or the flow rates tested (30-200 ml/min). There was a nominal drop in temperature as the power level was decreased (~2300 K at 30 W). The discharge is no longer sustained at lower power levels.

The most striking result of these measurements is the magnitude of the temperature, which is significantly less than the 3500-7500 K range reported for He MIP systems *(7)*. The observed differences are somewhat misleading since the HDD measurements were made in the secondary discharge region. Significant changes in excitation temperatures as a function of power levels, flow rates, and the cross section of plasma viewed have been reported for the MIP *(8)*. Many of the He transitions used in the MIP study were the same as those used in this investigation. Since excitation temperatures are characterized by the intensities of each He transition used, the population of the He excited states, in particular the metastable states, must be somewhat different in the secondary discharge region of the HDD than those associated with the primary discharge of the MIP where analyte excitation occurs.

Excitation Considerations. The exact mechanistic pathways through which the discharge is generated or atomic emission is produced in the HDD are by no means understood. A popular explanation for excitation of analyte species within the He MIP involves a sequential process, in which collisional energy transfer from He metastable species results in Penning ionization of the analyte element *(7,9)*. Such a process would explain the ability to observe ion emission from elements such as Cl and Br at trace levels in the MIP. Ion recombination with low energy electrons in turn will produce neutral excited atoms, from which characteristic emission can be observed.

One could rationalize that a given analyte molecule or some fragmented molecule could readily accept the discrete energies associated with the He metastable states. Any molecule will possess varying multitudes of electronic, vibrational, and rotational energies. The Penning ionization mechanism would also allow for excess energy to be imparted on ionized electrons as kinetic energy. Indeed, the energy associated with the He metastable states (19.5 eV and higher) creates a medium through which any element in the periodic table should be effectively excited.

The primary differences in the MIP and HDD are the excitation temperatures and the fact that insignificant ion emission is observed in the HDD for elements such as Cl and Br. Very little variation is observed in the intensities of neutral atom emissions relative to reported MIP values *(10)*. The relative absence of ion emission implies that if Penning ionization produces analyte ions, then the low energy electron population in the HDD must be substantially higher than the MIP, which would increase ion recombination and suppress ion emission.

The low excitation temperature measured for the HDD has been shown to be relatively independent of the flow rates studied. The best speculation is that the He metastable population in the HDD is higher than the MIP and less prone to collisional deactivation. This lack of deactivation may be due to the discharge being somewhat collapsed and not necessarily residing on the walls of the quartz tube. This may also account for the fact that no devitrification of the quartz walls has ever occurred.   Attenuated deactivation would also help explain the low background intensity associated with the HDD, which effectively enhances the signal to background ratios necessary to detect emission from analyte species at picogram levels. The low background also allows for the use of long slit heights (1.5 cm) and wide slit widths (typically 150 μm), which effectively covers a large area of the secondary discharge and minimizes observation height and spatial dependencies on the emission signals.

Although no concrete conclusions can be reached at this time, excitation and sensitivity to analyte emissions in the HDD is probably the result of higher low energy electron and He metastable atom populations relative to the MIP. Measurements of the electron density and temperatures, and the intensities of vacuum UV and IR emissions originating from the He metastable states may reinforce these conclusions.

Analytical Performance

Recent Advances. A review of the original HDD publication with respect to analytical figures of merit shows that the absolute limits of detection ranged from 0.5 pg for Hg to 50 pg for Si. The typical linear range of response was over three orders of magnitude. A significant drawback is the poor selectivity relative to the detection of non-analyte species at the analytical wavelengths used for element selective detection. This problem is particularly noticeable in the UV region of the spectrum where the selectivity ratios relative to octane were generally less than 100. Selectivity has been improved somewhat in the current instrument configuration due to better stray light rejection in the 0.5 m monochromator. Although background correction via oscillating refractor plates or photodiode arrays could be utilized, the potential variation in gas compositions or elemental compositions of interfering constituents in chromatographed samples can require extensive trial and error formulations or recipes to nullify correctly contributions to the element selective signal from non-analyte species.

Considerable efforts have recently been made to evaluate alternative wavelengths in the near IR which are less susceptible to the spectral interferences previously described. Volatile compounds containing the elements of interest were continuously purged at a constant rate through the secondary discharge by a helium carrier gas. Emission spectra were obtained from 600-1100 nm using a silicon photodiode as the detector. A typical wavelength scan obtained for sulfur emission lines is shown in Figure 3. All recorded intensities for observed emission lines were adjusted for the detector response and grating efficiency. The five most intense emission wavelengths for the elements investigated are given in Table II and normalized to the most intense emission for each element. A comparison to normalized intensities in the near IR for the MIP is shown for comparison. The observed intensities for the HDD and MIP are very similar with only a few exceptions.

The use of non-traditional near IR wavelengths for several of the elements is significant with respect to compromises between improved selectivity without background correction and limits of detection. Chromatograms from 1.0 µL injections of a mixture of thiophene and methyldisulfide at various concentrations in pentane are shown in Figure 4 with emission measured at the 921.23 nm line for sulfur. The limits of detection are markedly poorer than those obtained using the 182.0 nm analytical wavelength, however the selectivity over non-analyte species is enhanced considerably (>600).

A comparison of limits of detection and selectivity obtained in this laboratory at typical analytical wavelengths which provide the optimum sensitivity and those obtained at alternative near IR wavelengths is given in Table III for Cl, Br, I, S, and P. The best limits of detection and selectivities for these elements using the MIP at optimum

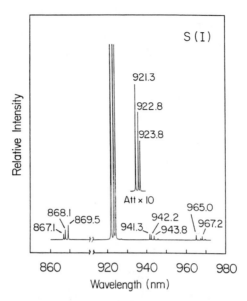

Figure 3. Near IR emissions observed from S(I) in the HDD.

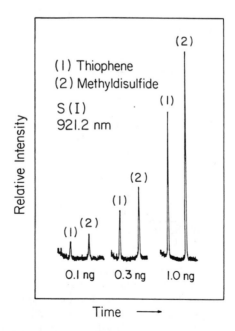

Figure 4. Sulfur selective chromatograms obtained at 921.2 nm for thiophene and methyldisulfide.

Table II. Relative Intensities of Emissions Observed in the Near IR for the HDD and MIP.

Wavelength(nm)		HDD	MIP[1]	Wavelength(nm)		HDD	MIP
N(I)	868.03	100	100	Cl(I)	912.11	100	100
	868.34	61	58		837.60	43	49
	862.92	35	27		894.80	33	37
	871.17	28	28		919.17	17	25
	870.33	26	7.3		958.48	16	13
P(I)	973.47	100	43	Br(I)	889.76	100	100
	979.68	29	100		827.24	55	27
	1059.69	24	6.2		882.53	35	40
	1045.59	23	1.6		863.87	31	30
	1008.42	16	23		926.54	29	55
S(I)	921.29	100	100	I(I)	973.17	100	61
	922.81	74	76		905.83	64	100
	923.75	47	47		965.31	44	34
	1045.56	9.5	5.0		1046.65	24	5.4
	1045.95	5.4	3.0		1023.88	8.9	3.5

[1]Reference 10.

Table III. Comparison of Limits of Detection and Selectivity at Selected Analytical Wavelengths.

	HDD			MIP[1]		
Element	λ(nm)	LOD(pg)	Selectivity	λ(nm)	LOD(pg)	Selectivity
Cl	837.5	3	1100	479.5	39	25000
	912.1	10	2500			
Br	827.2	8	160	478.6	75	19000
	892.0	15	880			
I	183.0	2	130	206.2	26	5000
	905.8	80	1700			
S	182.0	5	80	180.7	1.7	150000
	921.2	40	650			
P	253.5	30	60	177.5	1.5	25000
	973.5	1500	800			

[1]Reference 1 for Cl, Br, S, and P. Reference 11 for I.

analytical wavelengths is presented for comparison. All selectivities for the HDD were measured relative to octane as the non-analyte species. The selectivity for I, S, and P was improved substantially when using the near IR wavelengths. The marked improvement in Br selectivity implies that the wavelength normally used must coincide with some unknown spectral interferences. The obvious compromise is the poorer limits of detection, which is substantial for I, S, and P. The utilization of a more responsive detector in the 850-1100 nm region of the spectrum would certainly improve the limits of detection substantially. Although the limits of detection for the halogens are substantially better with the HDD, in all cases significant improvements in selectivity will have to be achieved for true element selective detection.

Applicable detection of C, H, N, and O has been prevented by excessively high background atomic emission from these species in the discharge support gas. Attempts to measure signals from GC eluents containing these elements actually suppresses the baseline signal. The HDD has the capability to determine empirical ratios of halogen species in compounds *(12)*, however determination of empirical formulas for compounds is not possible at this time.

Applications. The GC-HDD system in this laboratory has been applied to the determination of halogenated compounds containing Cl and Br. The HDD response was assumed to be based solely on the mass of the halogen eluting through the discharge and independent of compound structure or extent of halogen substitution. The same quantitative capabilities have been demonstrated for the MIP.

Individual polychlorinated biphenyl (PCB) congeners have been quantitated in standard solutions at ppm levels with an average error of less than 4% *(13)*. Hexachlorobenzene (HCB) was added as an internal standard in all the PCB mixtures to establish the relative peak area per unit concentration of Cl present from the HCB concentration. Identification of the congeners from retention times allowed for conversion of each PCB chromatographic peak area (detected as Cl emission) to PCB concentration based on the signal generated by the HCB. The HDD did not require any precalibration or response factor formulations which are necessary for common detectors such as the flame ionization and electron capture detectors.

The percent Cl in standard Aroclor samples, which contain a number of PCB congeners, could easily be determined by taking the sum of the entire Cl area generated by the PCBs relative to HCB without prior identification of the individual PCB congeners present. The most difficult problem associated with PCB quantitation was proper identification of the individual congeners (209 total) for relating the area generated from Cl emission to the concentration of the PCB. Significant progress has recently been made on compiling a complete retention index for all 209 congeners based on the correlation of a subset of 38 experimentally known congener retention times to those given in the literature for all 209 congeners obtained under different experimental conditions *(14)*. The accuracy of the HDD for PCB quantitation based on individual PCB congeners in Aroclor samples is demonstrated in Table IV.

Table IV. Average Compositions of Selected Aroclors Determined with the GC-HDD System.

Aroclor	Avg#Cl /Molecule		Weight %Cl		Avg Molec.Wt.	
	Expt	Lit[1]	Expt	Lit	Expt	Lit
1221	1.2	1.2	22.0	21	195.3	193.7
1232	2.0	2	31.7	32-33	224.9	223.0
1242	3.1	3	42.6	40-42	256.3	257.5
1248	3.9	4	48.8	48	286.5	291.9
1254	5.0	5	54.5	52-54	326.6	326.4

[1]Reference 15

Similar methods of quantitation have been applied to the determination of simple halogenated aliphatic and aromatic compounds (12). Absolute limits of detection ranged from 3 pg for chloroform and carbon tetrachloride to 29 pg for p-dibromobenzene. An average error of 1.5% was obtained when quantitating standard mixtures of compounds at concentrations of 1.0 and 10 µg/mL by using internal standards for calibration of the halogen area in a similar manner to the PCB work. Lower concentrations (<0.2 µg/mL) resulted in quantitation errors as high as 15%, however these errors appeared to be due more to inaccuracies in peak integrations at high amplifier gains.

The mass response feature of the HDD allowed for relatively simple quantitation of haloform species in drinking water samples through the addition of 2-chloro-1-bromopropane as an internal standard for calibration of both the Cl and Br response. Haloforms were isolated from 100 mL water samples by extraction into 15 mL of pentane, which was then concentrated to 1.0 mL with the addition of the internal standard for quantitation by GC-HDD. Relative limits of detection were <0.2 ppb for each compound.

Conclusions

A review of the criteria given in the beginning of this chapter for a successful element selective detector reveals that the HDD shares many of the attributes of the MIP with the exception of selectivity and lack of response to elements such as N and O, but the simplicity associated with the HDD warrants continued research and development. Areas of research which have virtually remained untouched since the initial conception include the overall design of the detector and means of more efficient coupling of electrical energy into the discharge. Refined gas purification, both in the GC and discharge gas lines, must be achieved before the detection of elements such as H, N, and O can be attempted. The reduction in contaminants should also improve the element selectivity and the potential to perform empirical formula determinations. The utilization of a photodiode array detector system would greatly enhance the overall capabilities of the HDD through simultaneous, multi-element detection and background correction for improved selectivity. The use of reagent gases (H_2, N_2) in the discharge should be evaluated for the enhancement of atomic emissions from elements such as O and S.

Research in the development of the HDD has been meager compared to the developed refinements in MIP systems. The evidence collected thus far for the viability of using the HDD as an element selective detector for GC certainly warrants continued research into instrument development and relevant applications.

Acknowledgments

This work has been supported in part by an ACS-PRF Grant (#17417-GB3) and a Thomas F. and Kate Miller Jeffress Memorial Research Grant (J-99, J-155).

Literature Cited

(1)   Quimby, B.D.; Sullivan, J.J. *Anal. Chem.* **1990**, *62*, 1027.
(2)   Sullivan, J.J.; Quimby, B.D. *Anal. Chem.* **1990**, *62*, 1034.
(3)   Olsen, K.B.; Griffen, J.W.; Matson, B.S.; Kiefer, T.C. Paper ANYL 195 ppresented at the 199th ACS National Meeting, Boston, MA, April 22-27, 1990.
(4)   D'Silva, A.P.; Zamzow, D.; Richard, J.J. Paper 903 presented at the 1989 Pittsburgh Conference and Exposition, Atlanta, GA, March 6-10, 1989.
(5)   Rice, G.W.; D'Silva, A.P.; Fassel, V.A. *Appl. Spectrosc.* **1985**, *40B*, 1573.
(6)   Ware, M.L. *The Development and Application of a Helium Discharge Detector for Gas Chromatography;* Masters Thesis, College of William and Mary, Williamsburg, VA, August, 1988.
(7)   Zander, A.T.; Hiefje, G.M. *Appl. Spectrosc.* **1981**, *35*, 357.
(8)   Workman, J.M.; Brown, P.G.; Miller, D.C.; Seliskar, C.J.; Caruso, J.A. *Appl. Spectrosc.* **1986**, *40*, 857.
(9)   Risby, T.H.; Talmi, Y. *CRC Crit. Rev. in Anal. Chem.* **1983**,*14*, 231.
(10) Pivonka, D.E.; Schliesman, A.J.J.; Fateley, W.G.; Fry, R.C. *Appl. Spectrosc.* **1986**, *40*, 766.
(11) Estes, S.A.; Uden, P.C.; Barnes, R.M. *Anal. Chem.* **1981**, *53*, 1829.
(12) Ryan, D.A.: Argentine, S.M.; Rice, G.W. *Anal. Chem.* **1990**, *62*, 1829.
(13) McAteer, P.J.; Ryerson, T.B.; Argentine, M.D.; Ware, M.L.; Rice, G.W. *Appl. Spectrosc.* **1988**, *42*, 586.
(14) Mullin, M.; Pochini, C.; McGrindle, S.; Romkes, S.; Safe, S.; Safe, L. *Environ. Sci. Tech.* **1984**, *18*, 468.
(15) Erickson, M.D. *Analytical Chemistry of PCBs*; Butterworth: Boston, MA, 1986, pp 18.

RECEIVED April 29, 1991

Chapter 13

Inductively Coupled Plasma Atomic Emission Spectrometry and Packed-Microcolumn Supercritical-Fluid Chromatography

Kiyokatsu Jinno, Hiroyuki Yoshida, Hideo Mae, and Chuzo Fujimoto

School of Materials Science, Toyohashi University of Technology, Toyohashi 441, Japan

Coupling of inductively coupled plasma atomic emission spectrometry and packed microcolumn supercritical fluid chromatography (SFC-ICP) has been investigated. The problems induced by this coupling have been evaluated and solved. The SFC-ICP has been proved as a highly informative hyphenated technique in SFC.

The use of supercritical fluid chromatography (SFC) is rapidly increasing as an alternative to gas chromatography (GC) and high performance liquid chromatography (HPLC) (1-3).Since supercritical fluids have properties intermediate between those of gases and liquids, the selectivity of SFC may resemble that of GC or HPLC depending on the volatility and solubility of the compounds to be separated and the operational temperature and pressure of the supercritical fluid. SFC may be used to separate multifunctional compounds which neither GC nor HPLC can separate.

Popular GC and HPLC detectors such as ultraviolet absorption (UV) (4-6) and flame ionization (FID) (7-10) have been used in SFC. In addition, information rich detectors such as mass spectrometers (MS) (11-13) and Fourier transform infrared spectrometers (FTIR) (14-16) have been coupled with SFC. However, none of these detectors provide elemental information on eluted compounds, and atomic emission spectrometric (AES) techniques are excellent candidates as information-rich SFC detectors. Plasma based detectors are promising because they have good detection limits, precision and/or accuracy, wide dynamic ranges, and capabilities for simultaneous multielement analysis. In addition, complete chromatographic resolution is not required because of their elemental selectivity. Microwave induced plasmas (MIP) have been shown to be useful element selective detectors for GC (17-19) and a similar approach can be applied to capillary SFC. Near-infrared detection combined with MIP is a powerful and high information content detector combination for capillary SFC of organo-chlorine and sulfur containing compounds (20,21). A radio frequency plasma detector showed high sensitivity for pesticide analysis with capillary SFC (22). Currently the well established inductively coupled plasma (ICP) provides an alternative mode for chromatographic detection. Various interfaces for HPLC or liquid flow injection

0097–6156/92/0479–0218$07.00/0

analysis (FIA) have been described. These involved either a spray-chamber system coupled with a cross-flow (*23*) or a concentric nebulizer (*24*), or alternatively "no-spray chamber" systems using either a cross-flow nebulizer (*25*) or a microconcentric nebulizer (*26*). These interfaces are also expected to work well with packed column SFC.

Recently Olesik and Olesik (*27*) reported introduction of a supercritical fluid into an ICP-AES instrument. A quartz capillary tube with a 1-15 μm terminal restriction was inserted into the central tube of a conventional ICP-AES torch less than 10 mm below the discharge region.

In this chapter, we describe an interface for coupling SFC with ICP-AES of different design from those developed for HPLC-ICP and FIA-ICP. Performance is investigated under "pseudo" and actual chromatographic conditions with 1mm i.d. and 0.5 mm i.d. packed capillary columns. The effects of the nebulizer gas and mobile phase flow-rates on the "pseudo " peak height, the linear dynamic range, the observation height, and the detection limits are studied with the interface. Supercritical carbon dioxide with and without polar solvents modifiers is evaluated and pressure programming is tested for ICP detection. Applications are reported and future potential is discussed.

EXPERIMENTAL SECTION

A schematic diagram of the instrumentation is shown in Figure 1 (*28*). The mobile phase was pure carbon dioxide or carbon dioxide modified with polar solvents. A reservoir was connected to a CCPD pump (Tosoh, Tokyo, Japan). In order to fill and pump the mobile phase smoothly, the pump head was cooled to ca. -5°C by a methanol/water mixture which was directed through a jacket by a circulating pump (Taiyo Kogyo, Tokyo, Japan). A piece of stainless-steel (SS) tubing (5 m x 0.8 mm i.d.) placed inside the jacket served as a precooling coil. Samples were injected with a Rheodyne Model 7520 injection valve (Cotati, CA, USA) with a 0.5 μL sample loop (at room temperature). All samples were dissolved in dichloromethane. The columns were a 25 cm x 1 mm i.d. SS tube and a 25 cm x 0.53 mm i.d. fused silica capillary packed with Shimpak Diol-150 (Shimadzu, Kyoto, Japan) in our laboratory by the slurry technique. The column temperature was maintained somewhat above the critical temperature of the mobile phase used. For optimization experiments, an empty pseudocolumn, a 25 cm x 0.8 mm i.d. SS tube, was used. The columns were placed in an oven kept at a constant temperature by a home-made thermostat. A 50 mm x 0.1 mm i.d. SS tube was used to connect the column to the injector. The column effluent was directed through a restrictor where direct nebulization into the plasma occured.

Carbon dioxide used had a purity of ca. 99.9 %. Methanol and acetonitrile (LC grade) were used as modifiers. Solvent mixing of the liquid carbon dioxide and the modifiers was accomplished "off-line" with a SS tank equipped with a stop valve and a short length of a 1/16 inch o.d. connecting tube. First, a given amount of modifier was introduced into the tank through the connecting tube. The liquid carbon dioxide was then pumped into the tank, which was cooled by ice-water, with the Tosoh CCPD pump. After disconnecting the tank from the pump, the two fluids were mixed by shaking the tank vigorously by hand. The tank was then connected to the CCPD pump to deliver the resultant mobile phase. The molar fraction of the modifier was calculated from the weights of the two fluids.

The column pressure was adjusted by setting the pressure of the pump, and thus the mobile phase flow rate was the independent variable. With this arrangement, the flow rate of the mobile phase through the column was calculated from the gas flow, which is the only measurable parameter.

A Model ICAP-1000 (Nippon Jarrell-Ash, Kyoto, Japan) inductively coupled

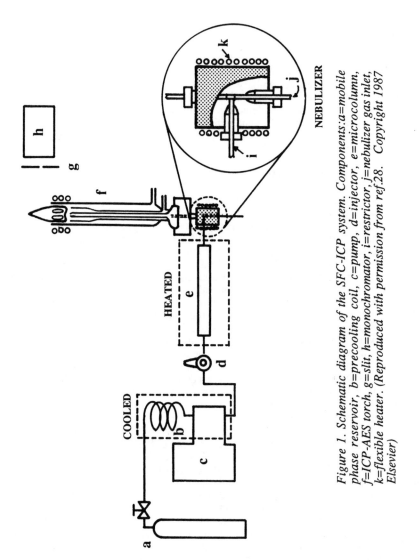

Figure 1. Schematic diagram of the SFC-ICP system. Components:a=mobile phase reservoir, b=precooling coil, c=pump, d=injector, e=microcolumn, f=ICP-AES torch, g=slit, h=monochromator, i=restrictor, j=nebulizer gas inlet, k=flexible heater. (Reproduced with permission from ref.28. Copyright 1987 Elsevier)

plasma spectrometer was employed, equipped with a 1 m Czerny-Turner monochromator. Analogue signals for single elements were recorded on a strip chart recorder. Operating conditions for the ICP are listed in Table I. The spray chamber supplied by the manufacturer was not used.

In the series coupling study of SFC and ICP-AES, a Uvidec 100 (Jasco, Tokyo, Japan) UV detector was employed. The UV detector was inserted between the outlet of the column and the restrictor. The stability study on organometallic complexes was performed by using a UV multichannel detector (Jasco, Multi-320 photodiode array detector). Data handling was with a NEC PC-9801 VX computer (Nippon Electric, Tokyo, Japan).

The amount of carbon dioxide which could be introduced into the ICP torch was measured and controlled using the modified experimental setup (29).

A flow restriction is needed to achieve supercritical conditions throughout the column. The restrictor employed is shown in Figure 2. It consisted of a glass capillary (35 mm x 50 μm i.d. x 0.5 mm o.d.) with a terminal restriction of ca. 5 μm. The restrictor was inserted into a SS tube (35.5 mm x 0.8 mm i.d. x 1/16 inch o.d.) to provide mechanical durability. The restrictor-tip design is similar to that described by Guthrie and Schwartz (30). The outlet of the restrictor was placed within a laboratory-made ICP cross-flow type nebulizer. Thus the restrictor can serve as a sample transfer line and sample introduction tube to the ICP nebulizer, in addition to its original function.

Two points appear crucial to the design of the SFC-ICP interface; minimization of extra column band-broadening and maximization of the efficiency with which effluent is introduced into the plasma. Because the interface or "nebulizer" investigated in this work has no spray-chamber, extra-column zone broadening is much less significant than with conventional ICP sample introduction systems. For the same reason, the sample introduction efficiency is higher than that obtained using conventional nebulizer introduction systems. A flow restrictor, a nebulizer gas tube (0.25 mm i.d., 1/16 inch o.d.) and a sample introduction tube (0.5 mm i.d., 1/16 inch o.d.) were installed to the three-way union. In order to prevent icing by the Joule-Thompson effect that occurs when the carbon dioxide expands to atmospheric pressure, the entire nebulizer was warmed with a flexible heater (k in Figure 1). More details are given below.

RESULTS AND DISCUSSION

System Performance. Effects on the emission signal intensity were investigated using the pseudocolumn under chromatographic conditions (supercritical FIA analysis). For this study, solutions of ferrocene (Fe, 120 ng/μL) were injected into the column and the emission signal intensity of the 259.94 nm Fe II emission line was observed with the ICP-AES system.

The effect of the nebulizer temperature was investigated with a column pressure of 160 atm. Emission intensities for multiple injections are illustrated in Figure 3. With the nebulizer at room temperature, "spiking" was observed in the signals and reproducibility was poor (Figure 3A). It is likely that the spiking is caused from intermittant freezing at the restrictor exit. In contrast, warming the nebulizer to 40°C gave reproducible peaks without spiking(Figure 3B), and a relative standard deviation of 2.1 % from six replicate injections. Reproducibility at 50°C and 60°C was similar to that at 40°C.

The nebulizer argon gas flow rate and the viewing height are also important parameters. Peak heights measured at various nebulizing gas flow rates and viewing heights are shown in Figure 4. The nebulizing gas flow producing the highest net emission intensity is the optimum. The data indicate that for greater viewing height the optimum gas flow rate is slightly shifted to the lower values. At very low

**Table I. ICP Conditions
for the Microcolumn SFC-ICP System**

Operating Conditions	
frequency	27.12 MHz
power	2.4 KW
Ar gas flow rate	
cooling gas	16 L/min
plasma gas	1 L/min
nebulizing gas	0.35 L/min
Viewing height	10 mm

To Column   To Nebulizer

*Figure 2. Schematic diagram of restrictor. Components:a=Pyrex glass capillary
(35 mm x 50 μm i.d., 0.5 mm o.d.) with an opening of ca. 5 μm, b=stainless
steel tube (34.5 mm x 0.8 mm i.d., 1/16 in.o.d.), c=epoxy resin.*

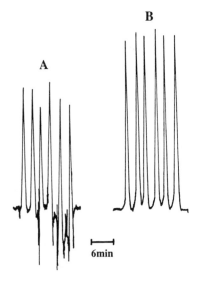

6min

*Figure 3. Reproducibility of emission signal intensity. Sample:0.5 μL of 0.4
μg/μL ferrocene in dichloromethane, nebulizer temperature:(A) room temperature,
(B) 40°C. Column pressure:160 atm, temperature: 40°C. (Reproduced with
permission from ref.28. Copyright 1987 Elsevier)*

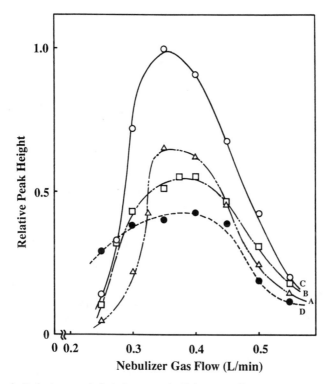

Figure 4. Relative peak height vs. nebulizing gas flow rate. A:5 mm, B:10 mm,C:15 mm, D:20 mm.

nebulizing gas flow rates, the amount of sample introduced is too small to give an adequate signal. At higher gas flow rates, the ICP plasma is cooled by the increase of the Ar gas loading and the conditions for dissociation and ionization become less favorable. Also, the residence times of ions in the plasma become shorter as the flow rate increases. A flow rate of 0.35 L/min Ar gas and a 10 mm viewing height provide the maximum signal intensity.

Column pressure is a most important parameter in SFC. Because the solubility of a solute is a function of the density of the supercritical mobile phase, the partition coefficients can be controlled by varying the pressure of the chromatographic system. In this study, the pressure was varied between 100 and 200 atm, and the heights and areas of the ferrocene peaks were measured. The results are shown in Figure 5. The carbon dioxide gas flow rate was calculated from the peak position by means of the FIA technique using the pseudocolumn. The flow rate of the mobile phase was estimated to range from 26 to 43 μL/min as column pressure measured from 100 to 200 atm. While the peak heights increase with the flow rate, the peak area is centered about a constant value. This data indicate that the ICP works as a mass-sensitive detector. However, it is noteworthy that the peak height at 10 mL/min gas flow is less than half of that observed at 20 mL/min. This is because, at low flow rates, the solute band is more spread out as it passes through the column.

Ferrocene solutions with concentrations ranging from 50 ng/μL to 2.5 μg/μL (i.e., Fe ranging from 7.5 ng to 375 ng) were injected into the pseudocolumn. The signal response was linear over about two orders of magnitude; the detection limit was about 7.5 ng Fe at S/N=4. The linear range would be expected to be further expanded if ferrocene were soluble in the sample solvent (dichloromethane) at concentrations above 2.5 μg/μL.

The system performance was evaluated for the 1 mm i.d. separation column. The effect of the nebulizing gas flow rate on the emission intensity of the 259.94 nm Fe line was measured at different viewing heights between 5 mm and 20 mm. The maximum signal intensity of the Fe component was observed at a viewing height of 10 mm and a nebulizing gas flow rate of 0.35 L/min. These results are in agreement with the values obtained when the pseudocolumn was used.

To demonstrate the potential of the ICP-AES detection system, a mixture of ferrocene derivatives was separated on the Shimpak 1 mm i.d. column at 40°C and 160 atm inlet pressure (flow rate was ca. 38 μL/min). The response monitored at the 259.94 nm Fe emission line is shown in Figure 6B. which shows well-defined peaks. A UV chromatogram for the same separation is shown for comparison (Figure 6A), chromatographic performance being almost identical.

Modifier Addition. The solvent strength of supercritical carbon dioxide changes markedly on increasing the density of the fluid. An alternative way to increase the solvent strength of the mobile phase is to add small amounts of miscible polar modifiers ; this offers operational flexibility since the modifier concentrations are easily varied.

Reproducibility was examined for methanol or acetonitrile modifiers (31), the 0.4 μg/μL ferrocene solution (Fe: 120 ng/μL) being successively injected into the pseudocolumn. Emission signals are shown in Figure 7, in which A and B indicate results with 8.0 mol % methanol and 7.3 mol % acetonitrile in carbon dioxide, respectively. The relative standard deviations of the data were 2.4 % and 3.3 %, respectively.

The effect of sample gas flow rate on the emission intensity was studied while varying the viewing height from 5 to 15 mm above the turn of the load coil. The mobile phase was 8.0 mol % methanol in carbon dioxide. In Figure 8, the maximum emission intensity is seen at a viewing height of 10 mm and nebulizing gas flow rate of 0.35 L/min, which are the same as those for pure carbon dioxide.

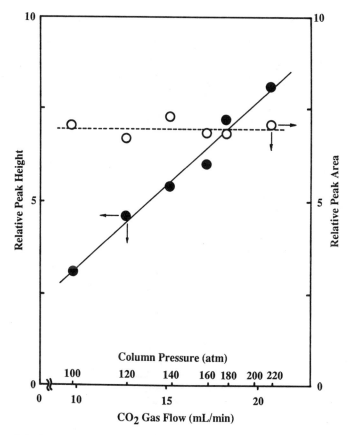

Figure 5. Relative peak height and peak area vs. column pressure. (Reproduced with permission from ref.28. Copyright 1987 Elsevier)

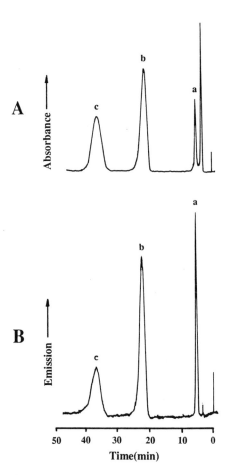

*Figure 6. Microcolumn separation of synthetic mixture of ferrocene. A :
chromatogram monitored by UV at 210 nm, B : chromatogram monitored by ICP
at 259.94 nm. Peak:a=ferrocene (0.4 µg/µL, 0.12 µg/µL Fe), b=acetylferrocene
(1.5 µg/µL, 0.37 µg/µL Fe), c=benzoylferrocene (1.5 µg/µL, 0.29 µg/µL Fe).*

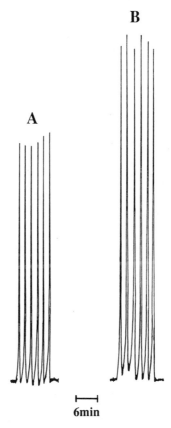

6min

Figure 7. Reproducibility of the emission signal intensity. (A) mobile phase, 8.0 mol % methanol in carbon dioxide. (B) mobile phase, 7.3 mol % acetonitrile in carbon dioxide. (Reproduced with permission from reference 31. Copyright 1989 Aster publishing)

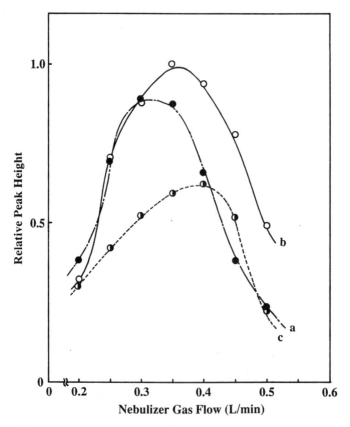

Figure 8. Relative peak height vs. nebulizing gas flow rate. Viewing height: a= 5 mm, b=10 mm, c=15 mm. (Reproduced with permission from reference 31. Copyright 1989 Aster publishing)

To study the effect of various modifiers on the emission intensity, carbon dioxide, 8 mol % methanol in carbon dioxide, and 7.3 mol % acetonitrile in carbon dioxide, were used. The relative peak heights in each case are similar and the greatest intensity was observed at a nebulizing gas flow rate of 0.35 L/min. The results suggest that modifiers such as methanol and acetonitrile give little or no change in the optimum nebulizing gas flow rate.

A mixture of three ferrocene derivatives was separated on the 1 mm i.d. Diol-150 column with 2 mol % methanol in carbon dioxide (Figure 9). The addition of small amounts of modifier results in improved peak shape, reduced retention times, and enhanced separation compared with the chromatographic behavior using pure carbon dioxide mobile phase (Figure 6B).

Loading Limit of Carbon Dioxide into ICP-AES. The advantage of varying modifiers in SFC-ICP is most significant when the analytes span a wide molecular weight/polarity range. Since the total effluent enters the ICP torch, quantitation should be straightforward. Also, it is expected that background signal changes due to organic modifiers will be small. However, knowledge of the loading limit of carbon dioxide into the SFC-ICP system is required for packed column SFC with conventional columns.

The loading limit has been examined to learn the highest mobile phase flow rate usable in this SFC-ICP system. Carbon dioxide was added to the ICP torch through the tee positioned behind the nebulizer. The ferrocene solution (Fe: 120 ng/μL) was injected into the pseudocolumn at an inlet column pressure of 160 atm, and column temperature of 40°C. The flow rate of gaseous carbon dioxide from the restrictor was ca. 26 mL/min. The emission intensity was constant in the gas flow range of 26-50 mL/min, at 55 mL/min it was 75 % of that in the 26-50 mL/min region, and the plasma was extinguished at 60 mL/min. Therefore, it was concluded that the upper loading limit of carbon dioxide is 50 mL/min, which corresponds to a supercritical fluid linear velocity of 4.0 mm/sec with a 1 mm i.d. microcolumn. It is noteworthy that ca. 40 min was required for the separation when the 1 mm i.d. column was used in conjunction with a linear velocity of 2.0 mm/sec as shown in Figure 6. For more rapid analysis the smaller i.d. columns are preferred. For example, when column diameter is reduced from 1 mm i.d. to 0.5 mm i.d., the linear velocity should be increased by a factor of 4. Hence the time required for separation should be decreased by a factor of 4. The loading limit of carbon dioxide into the ICP, 50 mL/min, corresponds to a supercritical fluid flow rate of 110 μL/min and a linear velocity of 16 mm/min with a 0.5 mm i.d. column.

Evaluation of 0.5 mm i.d. Column. The effects of flow rate of the nebulizing gas and column pressure on peak height, peak area, linear dynamic range, reproducibility and detection limit were studied for the 0.5 mm i.d. column.

To test reproducibility of the SFC-ICP system, 0.2 μg/μL ferrocene (Fe: 60 ng/μL) solution was injected into the column maintained at 160 atm. The response peaks obtained for the repeated injections gave a 1.8 % relative standard deviation of the peak height. The lowest detectable concentration at a 0.5 μL injection was 20 ng/μL for ferrocene (Fe: 6 ng/μL) at a signal-to-noise ratio of 4.

The effect of the nebulizing gas flow rate on the emission intensity observed in the SFC-ICP system was investigated using the 0.5 mm i.d. diol column. The nebulizing gas flow rate was varied from 0.2 L/min to 0.5 L/min. The viewing height above the load coil was varied from 5 to 20 mm. The maximum emission intensity was obtained at the nebulizing gas flow rate of 0.35 mL/min and the viewing height of 10 mm. These values are the same as those obtained using the pseudocolumn and the 1 mm i.d. column. The effect of the column pressure on the emission intensity was also investigated and the results are shown in Figure 10.

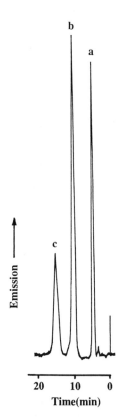

Figure 9. Microcolumn SFC-ICP chromatogram of a synthetic mixture of ferrocene. Peaks:a=ferrocene(0.2 µg/µL, 60 ng/µL Fe), b=acetylferrocene(0.5 µg/µL, 0.12 µg/µL Fe), c=benzoylferrocene(0.6 µg/µL, 0.12 µg/µL Fe). (Reproduced with permission from reference 31. Copyright 1989 Aster publishing)

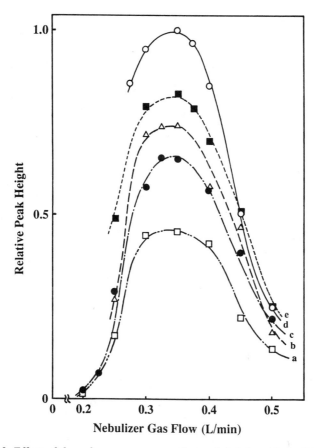

Figure 10. Effect of the column pressure on the peak height with the 0.5 mm i.d. column. Column pressure:a=120 atm, b=140 atm, c=160 atm, d=180 atm, e=200 atm. (Reproduced with permission from reference 29. Copyright 1990 Aster publishing)

The column pressure and the nebulizing gas flow rate were varied from 120 atm to 200 atm and from 0.2 L/min to 0.5 L/min, respectively. The pressure over the column was adjusted by controlling the pressure of the pump. When the pressure was varied from 120 atm to 200 atm, the gas flow rate, which had been monitored independently at the exit of the restrictor, changed from 22 mL/min to 30 mL/min. This means that the mobile phase flow rate changed from 57 μL/min to 78 μL/min. It has been found that the maximum peak intensity is obtained using the nebulizing gas flow rate of 0.35 L/min, regardless of column pressure.

As discussed above, the ICP detector is mass sensitive, even if various parameters of SFC such as the column pressure, the column temperature and the modifier concentration, are changed. To investigate the effect of such parameters on the peak area, a 0.1 μg/μL ferrocene solution (Fe 30 ng/μL) was injected into the diol column and the column pressure was changed between 120 atm and 200 atm. The peak area was calculated from the resultant peaks. The results are summarized in Figure 11(A). Then, the column temperature was changed between 40°C and 65°C while maintaining the column pressure at 160 atm, and the results are shown in Figure 11(B). In Figure 11(C), the effect of the modifier concentration (methanol in this case) on the peak area is also demonstrated, where methanol concentration was increased up to 15 mol %. The ICP detector serves as the mass-sensitive detector for SFC.

Linear dynamic range was also examined. Ferrocene solutions containing from 10 ng to 1.25 μg were injected into the 0.5 mm i.d. diol column at a pressure of 160 atm. The peak height increases linearly from the detection limit, 10 ng ferrocene (Fe: 3 ng) to 2.5 μg ferrocene (Fe: 375 ng).

To demonstrate the potential of the 0.5 mm i.d. column a mixture of ferrocene derivatives was separated on the diol column (0.5 mm i.d. x 250 mm length) with carbon dioxide as the mobile phase (31). The separation is completed within 8 min, much more rapidly than with the 1 mm i.d. column. The use of 2 mol % methanol modifier and the 0.5 mm i.d. column enables the analysis to be completed within 4 minutes with same loss in resolution. There is no doubt that higher velocities of the mobile phase can enhance the absolute detection limit of the ICP detector.

In SFC, pressure is a very important parameter because of its ability to change the solvating properties of the mobile phase significantly and thus, pressure programming is a means of changing and extending the effective eluting power of the supercritical fluid mobile phase. Separation of four ferrocene derivatives was performed with step-wise pressure programming; column pressure was maintained at 100 atm and raised to 160 atm after 4 min. The chromatogram obtained is shown in Figure 12. At 100 atm, ferrocene and n-butylferrocene were eluted separately but acetylferrocene and benzoylferrocene were not eluted from the column.

Separation of Organometallic Compounds. SFC-ICP detection of metal acetylacetone (acac) complexes was also studied. Wenclawiak et al. (32) have described the separation of metal acac complexes by SFC. They used a conventional LC column packed with silica or octyl bonded silica and a binary solvent mixture of up to 30 mol % methanol or ethanol in carbon dioxide. Taylor et al.(33) reported the SFC separation of metal acac complexes, where conventional LC columns were used with carbon dioxide modified with 20 wt % methanol. They noted that two different chromatographic behaviors were observed, dependent on the specific metal coordinated to the acetylacetone ligand. Inert complexes probably associate through the outer coordination sphere, whereas labile complexes may interact with the stationary phase directly through the metal, after loss of one of the peripheral ligands. However, because they used a UV detector, the retention time was the only available information. Speciation of the eluates is possible using element-specific detection. To demonstrate this approach, SFC, UV, and ICP were coupled in series.

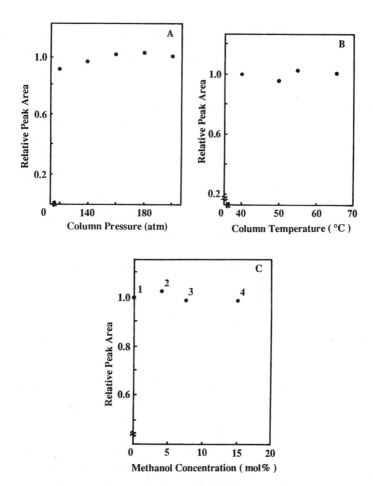

Figure 11. A:Effect of the column pressure on the relative peak area at a column temperature of 40 °C. B:Effect of the column temperature on the relative peak area at a column pressure of 160 atm. C:Effect of methanol modifier concentration on the relative peak area at a pressure of 160 atm and the temperatures of 55 °C for 1,2 and 3, 65 °C for 4.

The metal acac complexes were injected into the column. A typical chromatogram for chromium and cobalt is shown in Figure 13. Other metal complexes, Al(acac)$_3$, Cu(acac)$_2$, Fe(acac)$_2$, Mg(acac)$_2$, Mn(acac)$_2$ and Zn(acac)$_2$ showed considerable tailing in the UV chromatograms, but no peaks in the ICP chromatograms. The same results were observed when a solution of acetylacetone was injected into the column. These results suggest that the tailing peaks are due to the ligand liberated from the complex. Thus, it appears that Co(acac)$_2$ and Cr(acac)$_3$ are stable while the other metal acac complexes investigated decomposed under the chromatographic conditions used.

In order to determine the stability of various acac complexes under supercritical conditions, UV spectra were measured using FIA combined with the UV multichannel detector (*34*). Mobile phases in the FIA measurements were dichloromethane(A), supercritical carbon dioxide (B, 140 atm, 50°C) and carbon dioxide modified with 5 mol % methanol (C, 140 atm, 50°C). Some selected spectra are shown in Figure 14. Comparing UV spectra of acac complexes to those of the acac ligand under these three conditions, shows differences among the spectra which reflect the stability of the complexes.

From these results, one can obtain the following information about the stability of acac complexes. In the dichloromethane environment only Co(acac)$_2$ is unstable,but in the supercritical state all acac complexes except Al(acac)$_3$, Co(acac)$_3$, Cr(acac)$_3$ and Cu(acac)$_2$ are unstable. Those unstable acac complexes are probably decomposed at the high temperature of the supercritical measurements.

The ICP detector was then connected to the supercritical FIA system to determine whether the components which showed different UV spectra from those of the ligand were correctly assigned as the desired acac complexes. Emission signals were observed for Al, Co(III), Cr and Cu complexes. Other complexes which appeared to be unstable were also examined but signals of the metal species were not observed. These experiments served to confirm that the conclusions from the results of the supercritical FIA-UV multichannel detector system are correct.

The separation for those four stable complexes was examined with the SFC-UV-ICP system (Figure 15). Three main peaks appeared, although four components were injected into the system. Spectra of peaks a and b were similar to those of Cr(acac)$_3$ and Co(acac)$_3$, respectively. Using ICP detection, two signals were obtained by monitoring the emission lines at 267.7 nm for Cr and 228.6 nm for Co, retention times being consistent with peaks in the UV chromatogram. Therefore, one can assign the peaks to Cr and Co complexes. Since the UV spectrum of the peak at 1.8 min in the UV chromatogram is similar to that of the ligand and no signals were observed at this retention time with ICP detection, it is clear that Al(acac)$_3$ and Cu(acac)$_2$ were not eluted as the complexes. When each Al and Cu complex was injected into the system, it gave the peak at the retention time of the acac ligand and very similar UV spectra. The results suggest that Al and Cu complexes decomposed in the SFC system, and the metals are retained. A notable difference between supercritical FIA (where Al and Cu complexes gave clear signals in UV and ICP detection) and SFC, is the separation column. Interaction of the complexes with the column packing materials may cause their decomposition under the SFC conditions.

Acac was added to the carbon dioxide / methanol mobile phase, it being expected that its presence would prevent liberation of the ligand from the complex. A mobile phase of 0.1 mol % acac and 4 mol % methanol in carbon dioxide was used, UV and ICP chromatograms of those complexes being shown in Figure 16. Peaks of each component are seen in the ICP chromatograms, while two peaks appeared in the UV chromatogram. UV spectra could not be obtained because of the high background caused by the absorption of the acac ligand, but the ICP detection indicates that the peak at 8 min in the UV chromatogram contains Al, Cr and Cu

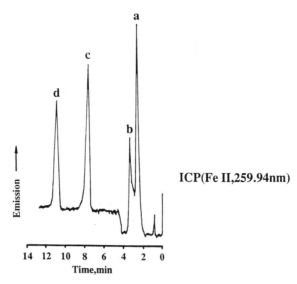

Figure 12. Stepwise pressure gradient separation of a synthetic mixture of ferrocenes. Peaks:a=ferrocene (0.2 µg/µL, 60 ng/µL Fe),b=n-butylferrocene (0.3 µg/µL,90ng/µLFe),c=acetylferrocene(0.5µg/µL,122ng/µLFe),d=benzoylferrocen e(0.30 µg/µL, 57 ng/µL Fe). (Reproduced with permission from reference 29. Copyright 1990 Aster publishing)

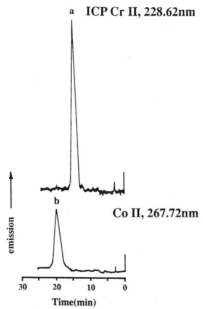

Figure 13. Element-selective detection of metal acac compounds. Peaks: a=Cr(acac)₃(0.44 µg/µL Cr), b=Co(acac)₃(0.33 µg/µL Co). Column:1 mm i.d., mobile phase:2 mol % methanol in carbon dioxide, pressure:160 atm, temperature:40°C. (Reproduced with permission from reference 28. Copyright 1990 Aster publishing)

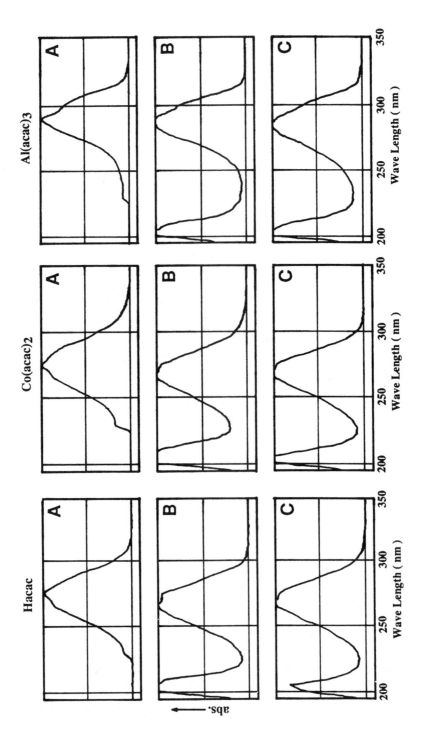

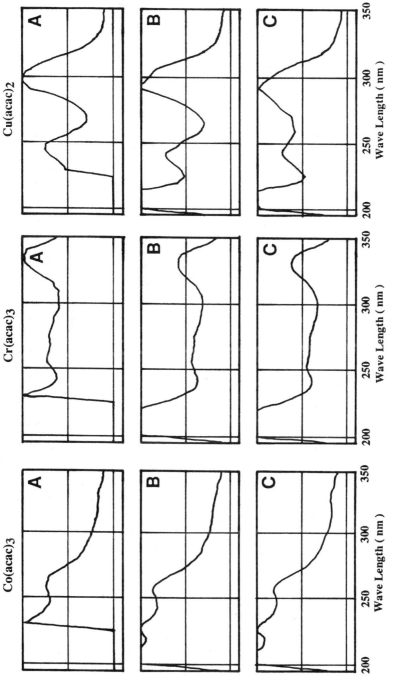

Figure 14. Typical UV spectra of the acac ligand,Co(acac)$_2$,Al(acac)$_3$,Co(acac)$_3$, Cr(acac)$_3$ and Cu(acac)$_2$. A:in dichloromethane, B:supercritical carbon dioxide, C:5 mol % methanol with supercritical carbon dioxide.

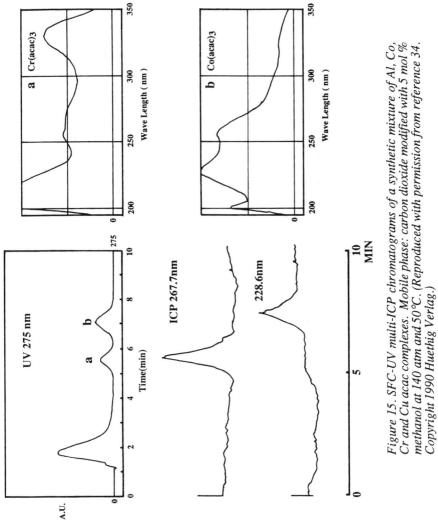

Figure 15. SFC-UV multi-ICP chromatograms of a synthetic mixture of Al, Co, Cr and Cu acac complexes. Mobile phase: carbon dioxide modified with 5 mol % methanol at 140 atm and 50°C. (Reproduced with permission from reference 34. Copyright 1990 Huethig Verlag.)

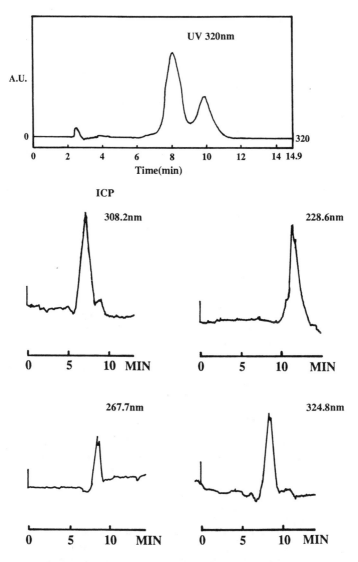

Figure 16. SFC-UV multi-ICP chromatograms of a synthetic mixture of Al, Co, Cr and Cu acac complexes. Mobile phase: carbon dioxide modified with acac 0.1 mol % and methanol 4.3 mol % at 140 atm and 50°C. (Reproduced with permission from reference 34. Copyright 1990 Huethig Verlag.)

complexes in this sequence and the peak at 10 min arises from the Co complex. It is clear from this experiment that Al and Cu complexes are stable in the presence of acac ligand in the mobile phase. Unfortunately, the other labile complexes which are unstable in supercritical conditions were not detected with either ICP or UV detectors with these SFC conditions.

CONCLUSION

The SFC-ICP system is an effective tool for speciation of metal-containing compounds, even when the modifiers are added to the supercritical carbon dioxide mobile phase to control the retention of components. An advantage results from the significantly reduced flow rates of microcolumns compared with those of conventional columns, due to the reduced amount of effluents which enter the ICP. In SFC, two basic approaches are used most often to control retention: pressure programming and modifier addition. Our results have clearly shown that for the SFC-ICP combination, both methods can be useful.

The future of ICP detection for SFC will be as a supplementary technique, rather than in direct competition with GC-ICP, because different types of compounds can be separated by SFC and GC. However, information gathered from ICP detection may yield a more effective way to perform elemental analysis of separated components when the detection region is expanded to the near infrared region.

REFERENCES

(1)   Novotny, M.; Springston, S.R.; Peaden, P.A.; Fjeldstad, J.C. and Lee, M.L. *Anal.Chem.* **1981,** *53*, 407A.
(2)   Fjeldstad, J.C. and Lee, M.L. *Anal.Chem.* **1984,** *56*, 619A.
(3)   Goew, T.H. and Jentoft, R.E. *J.Chromatogr.* **1972,** *68*, 303.
(4)   Sie, S.T. and Rijnders, G.W.A. *Sep.Sci.* **1967,** *2*, 726.
(5)   Novotny, M.; Bertsch,W. and Zlatkis, A. *J.Chromatogr.* **1971,** *61*, 17.
(6)   Jinno, K.; Hoshino, T.; Hondo, T.; Saito, M. and Senda, M. *Anal.Chem.* **1986,** *58*, 2696.
(7)   Wright , B.W. and Smith, R.D.*Chromatographia* **1984,** *18*, 542.
(8)   Kong, R.C.; Woolley, C.W.; Fields, S.M. and Lee, M.L. *Chromatographia* **1984,** *18*, 362.
(9)   Rokushika, S.; Naikwadi, K.P.; Jadhav, A.L. and Hatano, H. *Chromatographia* **1986,** *22*, 209.
(10) Hawthorne, S.B. and Miller, D.J. *J.Chromatogr.Sci.* **1986,** *24*, 258.
(11) Smith, R.D.; Udseth, H.R. and Wright, B.W. *J.Chromatogr.Sci.* **1986,** *24*, 258.
(12) Smith, R.D.; Kalinoski, H.T.; Udseth, H.R. and Wright, W.R. *Anal.Chem.* **1984,** *56*, 2476.
(13) Crowther, J.B. and Henion, J.D. *Anal.Chem.* **1985,** *57*, 2711.
(14) Shafer, K.H. and Griffiths, P.R. *Anal.Chem.* **1983,** *55*, 1939.
(15) French, S.B. and Novotny, M. *Anal.Chem.* **1986,** *58*, 164.
(16) Johnson, C.C.; Jordan, J.W. and Taylor, L.T. *Chromatographia* **1985,** *20*, 717.
(17) Beenakker, C.I.M. *Spectrochim.Acta* **1977,** *PartB*, *32B*, 173.
(18) Risby, T.H. and Talmi, Y. *CRC Crit.Rev.Anal.Chem.* **1983,** *14*, 231.
(19) Sullivan, J.J. and Quimby, B.D. *HRC.,* **1989,** *12*, 282.
(20) Galante, L.J.; Selby, H.; Luffer, D.R.; Hieftje, G.M. and Novotny, M. *Anal.Chem.,* **1988,** *60*, 1370.

(21) Luffer, D.R.; Galante, L.J.; David, P.A.; Novotny, M. and Hieftje, G.M. *Anal.Chem.* **1988,** *60*, 1365.
(22) Shelton, R.J. Jr.; Farnsworth, P.B.; Markides, K.E. and Lee, M.L. *Anal.Chem.* **1989,** *61*, 1815.
(23) Fraley, D.M.; Yates, D.A.; Manahan, S.E.; Stalling, D. and Petty, J. *Appl.Spectrosc.* **1981,** *35*, 525.
(24) Carnahan, J.W.; Mulligan, K.J. and Caruso, J.A. *Anal.Chim.Acta* **1981,** *130*, 227.
(25) Jinno, K.; Nakanishi, S. and Fujimoto, C. *Anal.Chem.* **1985,** *57*, 2229.
(26) Laurence, K.E.; Rice, G.W. and Fassel, V.A. *Anal.Chem.* **1984,** *56*, 289.
(27) Olesik, J.W. and Olesik, S.V. *Anal.Chem.*, **1987,** *59*, 796.
(28) Fujimoto, C.; Yoshida, H. and Jinno, K. *J.Chromatogr.* **1987,** *411*, 213.
(29) Jinno, K.;Yoshida, H. and Fujimoto, C.*J.Microcolumn Sep.*, **1990,** *2*, 146.
(30) Guthrie, E.J. and Schwartz, H.E. *J.Chromatogr.Sci.* **1986,** *24*, 236.
(31) Fujimoto, C.; Yoshida, H. and Jinno, K. *J.Microcolumn Sep.*, **1989,** *1*, 19.
(32) Wenclawiak, B. and Bickmann, F. *Z.Anal.Chem.* **1987,** *320*, 261.
(33) Ashraf-Khorassani, M. Hellgeth, J.W. and Taylor, L.T. *Anal.Chem.* **1987,** *59*, 2077.
(34) Jinno, K.; Mae, H. and Fujimoto, C. *HRC.* **1990,** *13*, 13.

RECEIVED April 11, 1991

Chapter 14

Helium High-Efficiency Microwave-Induced Plasma as an Element-Selective Detector for Packed-Column Supercritical-Fluid Chromatography

Gary L. Long, Curtis B. Motley[1], and Larry D. Perkins[1]

Department of Chemistry, Virginia Polytechnic Institute and State University, Blacksburg, VA 24061–0212

The development and use of a helium high efficiency microwave induced plasma (He HEMIP) for the spectrometric determination of metals and nonmetals in the effluent of a packed column supercritical fluid chromatograph (SFC) is discussed. Studies concerning the operation, the analytical utility and the effect of introducing CO_2 at packed column flows rates on the plasma are presented. The methods used to interface the SFC to the HEMIP are described. The use of pressure programming of the SFC pump is shown to have little effect on the operational characteristics of the HEMIP. The detector possesses linearity over 5 orders of concentrative magnitude. Minimum detectable quantities for the nonmetals range from sub to low ng/mL.

Over the past three years, various efforts to develop an element specific detector for supercritical fluid chromatography, SFC, using plasma emission spectrometry have been reported in the literature. Plasmas that have been used are the inductively coupled plasma, ICP (1,2), a radio frequency discharge (3), and several forms of the microwave induced plasma [4-7]. Regardless of the plasma used, the source must fulfill certain basic criteria in order to function as an acceptable elemental detector for SFC. The plasma must be able to promote the transitions of interest (metals as well as nonmetals) and be unaffected in terms of operating characteristics by the introduction of CO_2 and modifying agents into the plasma. These operating characteristics are the excitation temperature, electron number density and stability of the plasma.

[1]Current address: Procter and Gamble Company, Miami Valley Laboratories, Cincinnati, OH 45239–8707

0097–6156/92/0479–0242$06.00/0

The latter criterion of being unaffected by the introduction of CO_2 has been shown to be the most difficult task for a plasma to fulfill. The CO_2 flow rates used in SFC can range from several mL/min to over a hundred mL/min depending on the type of column and the pump pressure used. Packed column flow rates are in the order of 60-120 mL/min of CO_2 while capillary column flows are a factor of 10 less. The introduction of carbon containing gases into Ar plasmas (ICP and MIP) has been shown to affect the excitation properties of the plasma adversely [5,8,9]. Those employing ICP for SFC detection have noted a degradation in plasma energy when CO_2 is introduced into the plasma at capillary column flow rates [10]. Additionally, the limited amount of energy possessed by the Ar discharge is inadequate to excite nonmetal species.

To overcome the shortcomings of the Ar plasma as an elemental detector for SFC, helium based plasmas have been employed [3,4,6,7,11-14]. Although He possesses a higher excitation energy and should result in a more energetic plasma than Ar, several He based plasmas have been reported to experience a substantial loss in plasma energy while using high CO_2 pressures in capillary column SFC [4,6,7]. Packed column SFC, with even higher CO_2 flow rates, has not been explored with these plasmas.

We believe that a critical factor in the operation of a He plasma as an SFC detector is the maintenance of effective power transfer from the generator to the resonant cavity for the various CO_2 flow rates encountered in the chromatographic experiment. For this reason we have been investigating a highly efficient microwave induced plasma, HEMIP, as an element specific detector for packed column SFC. This cavity has been developed by Matus, Boss and Riddle [15] and subject to considerable study and further modifications by Boss [15,16] and Long [5,11,12,17-19]. The HEMIP cavity may be critically tuned and matched to the generator to compensate for the effect of added CO_2 to the plasma [12,19]. Normally, the power transfer efficiency to the HEMIP cavity is in excess of 95% [16]. The use of He in the HEMIP is now shown not to be affected by the introduction of CO_2, even when this gas comprises 20% of the total plasma gas flow. The HEMIP is the only plasma based detector that has been reported to be capable of functioning as an adequate elemental detector for packed column SFC [12].

In this paper, the interfacing of the HEMIP to a packed column SFC is described. The effect of the introduction of supercritical CO_2 on the plasma excitation temperature, electron number density, plasma geometry and cavity coupling will be reported for a HEMIP using He. Studies on the effect of plasma gas flow rates, pressure programming, and applied power is presented. Minimal Detectable Quantity values for select nonmetals and metals are given.

Experimental Considerations

Reagents. All chemicals used were analytical reagent grade. Water was distilled and deionized. Stock solutions of all aqueous metals were purchased as 1000 ppm (Buck Scientific, Inc.) or prepared following standard procedures [20]. Iron for organic solvent determinations, was prepared by dissolving ferrocene in xylenes to obtain a 1000 ppm solution. Volumetric dilutions of these solutions were performed to obtained desired concentrations.

The plasma gas used was Airco analytical-grade helium. Carbon dioxide for the mobile phase of the SFC was purchased as chromatographic grade (Scott Specialty Gases) with a helium headspace of 1500 psi and dip tube. All samples were prefiltered through 0.5 µm Teflon filters (Fisher Scientific).

Optical System. Emission from the plasma was focused onto the entrance slit of the monochromator (0.35 m, Heath Corp.; 0.25 m, PTI Co. for nonmetal determinations) using a f/3 lens for 1:1 imaging. Current signals from the photomultiplier tube were converted to voltage using a current to voltage converter (Heath Corp, or PTI Co.) in order to make the final signal compatible with the input of the 12-bit analog-to-digital converter (ADALAB, Pittsburgh, PA) on an Apple IIe computer or the input of a strip chart recorder.

Two different viewing geometries were used for collection of data with the He HEMIP. Studies of metal atomic emission were conducted with the plasma in the radial position (side-on), unless otherwise stated. An axial observation position (end-on) was used for all nonmetal studies [21].

In order to use the emission lines below 190 nm, the monochromator and optical path were purged with Ar gas. The laboratory constructed purge system consisted of pyrex glass and rubber o-rings connected to an aluminum plate attached to the entrance slits of the monochromator. A f/3 lens was placed mid-way in the purge tube to provide 1:1 imaging of the axial emission of the plasma onto the entrance slits of the monochromator.

SFC Instrumentation. A Suprex 200/A SFC pump was used in this work. The CO_2 was prefiltered through a 0.5 µm stainless-steel frit filter prior to filling the SFC pump which has a 200 mL capacity. DeltaBond phenyl and DeltaBond cyano columns (Keystone Scientific, PA) used in these studies have an inner diameter of 1 mm and were 10 cm long with 5 µm particles. Injections were with a high-performance liquid chromatography Rheodyne Model 7250 valve (Cotati, CA) equipped with a 0.5 µL injection rotor. In order to eliminate band broadening at the injector point, the head of the column was placed as close as possible to the injection rotor. All injections were performed in a splitless manner. The entire injection apparatus was mounted outside the oven, to allow injections to be made at room temperature.

In order that elemental detection at atmospheric pressure could be accomplished, a tapered restrictor was employed [22]. The tapered restrictors were produced in the laboratory and were fashioned from 50 µm fused silica to an orifice approximately 10 µm in diameter.

Microwave Cavity. The microwave cavity used is diagrammed in Figure 1. The design is a modified version of that described by Boss, Matus, and Riddle [15]. Modifications included a change in size and reconstruction of the probe translation stage. The cylindrical cavity was machined from a 19 mm thick sheet of oxygen-free high conductivity copper (OFHC). The internal diameter and depth of the cavity were fixed at 96.0 mm and 10.0 mm, respectively. Three 8 mm diameter quartz rods extended into the cavity from the side wall. Movement of the quartz rods provided tuning of the cavity resonant frequency to the generator frequency, 2.45 GHz.

A removable cavity lid was machined from a 5 mm OFHC sheet. An 8 mm hole was drilled into the center of the lid to facilitate insertion of the plasma discharge tube. A 5 x 40 mm radial slot was milled along a radius of the circular lid. This allowed lateral movement of the capacitive antenna coupling probe.

The capacitive antenna coupling probe was used for proper coupling between the generator and the load (plasma). The capacitance of the probe was adjusted by varying the surface area at the end of the probe. Coupling was further adjusted by sliding the probe along the radial face of the cavity to match the impedence of the cavity (capacive and inductive) with the impedance of the generator.

Antenna probes were fashioned from UG 58 A/U type N coxial connectors, as described by Boss [15,16]. A 10 gauge copper wire was soldered to the center post of the N connector. A 16 mm diameter stainless steel disk was soldered to the end of the copper wire. Sheet metal shims were used to set the probe penetration depth at 96%. The probe translation mechanism was constructed from brass.

Microwave Torch. A centered plasma was produced using a tangential torch modified from a design by Deutsch and Hieftje [23]. The torch used in all studies consisted of two concentric quartz tubes (see Figure 2). The dimensions of the outer tube were 6 mm i. d., 8 mm o. d. and 9 cm long. The inner tube was 1.5 mm i. d., 2 mm o. d. and 2.5 cm long. For all experiments, the He flow was introduced tangentially into the outer tube, allowing for a homogeneous mixture to form between the SFC effluent and plasma gas before ignition.

Plasma Ignition. In operation, the He HEMIP is self-igniting. On those occasions when the plasma did not ignite the following procedure was employed. The circular probe was placed near the center of the cavity. The plasma gas was allowed to flow through the side inlet of the torch (i.e., 1-2 L/min). All gas flows were metered with 150 mm rotameter (Air Products). Microwave power levels of up to 150 W were supplied to the cavities by either a Holiday Industries HI-2450 generator or a Micro-Now 420 generator. The quartz tuning rods were adjusted to obtain a minimum reflected power which was monitored by a directional coupler (Micro-Now Industries) and a power meter (Bird Electronics). The plasma was ignited with a small tungsten wire attached to a rubber policeman, whereby inductive heating of the wire caused a seeding of the He gas. If the plasma did not ignite, the probe was moved toward the walls of the cavity, and the above process was repeated. Once the plasma ignited, the reflected power was minimized by adjustments of the antenna probe and quartz rods. The flow was reduced to 1 L/min.

Aqueous Sample Introduction. For diagnostic studies, sample solutions were introduced into the plasma with a typical ICP-AES Meinhard pneumatic nebulizer/Scott spray chamber system. The optimum operational parameters for all studies involving aqueous sample introduction into the He plasma were 150 W applied power, 0 W reflected power, a He flow rate of 1 L/min, a sample uptake rate of 0.46 mL/min, and a viewing height of 3 mm above the face of the cavity for radial studies.

SFC Sample Introduction. Samples were introduced into the plasma by inserting the fused silica restrictor into the plasma torch. The first method, termed central introduction, involved sliding the restrictor into the inner tube of the plasma torch. The tip of the tapered restrictor was allowed to protrude from the end of the inner plasma torch tube to a distance of 1 cm from the plasma discharge. The restrictor was inserted through a GC septum in order to hold it in place and seal the torch from leaks. The septum was pressed into the plasma torch tube. The second method involved metering the SFC effluent into the side arm of the plasma torch by a Pyrex "T", in which the decompressing SFC effluent was mixed with the plasma gas. A GC septum was used to hold the restrictor in place. The plasma gas flow was at a right angle to the restrictor tip to prevent the nucleation of CO_2 as it decompressed.

Plasma Stability. The He HEMIP was stable over all operating periods used to gather data (30 min to 10 hrs). Additional tuning was not required during these time periods. The temperature of the cavity, cables, and connectors did not exceed 35°C.

Minimal Detectable Quantity. The term minimal detectable quantity (MDQ) was calculated according to the guideline of Hartmann for GC detectors [24]. It is defined as $MDQ = 2*N/S$, where N is the noise level and S is the analytical sensitivity.

Diagnostic Parameters. The electronic excitation temperature of the He HEMIP was determined from the spectral emission intensities of helium and iron as thermometric species [25]. Helium atom lines in the region of 380-502 nm were measured. An iron solution of 1000 ppm was introduced into the plasma and the relative intensities of the iron atom lines in the spectral region of 370-390 nm were measured. A comparison of these intensities yielded an excitation temperature based on a Bolzmann distribution of the excited state species using an equation described by Blades et al. [26].

The spatial ionization temperature, T_{ion}, was determined from the relative emission intensities of the cadmium atom (228.8 nm) and cadmium ion (226.5 nm) lines using the same position and resolution as the measurements for the excitation temperature, and was calculated using the Saha-Eggert relationship [26].

The electron number density of plasmas were determined from the width of the Stark broadened hydrogen-beta line (486.13 nm) at half maximum using the method of Griem [27].

Results and Discussion

The HEMIP has been utilized in our laboratory as a source for the atomic spectrometric determination of nonmetals and metals in packed column SFC chromatographic effluent as well as with direct aqueous sample introduction from a pneumatic nebulizer. The He plasma has been characterized using these sample introduction methods. In the following sections, discussion will be presented on the interfacing of the SFC to the HEMIP cavity, diagnostic studies on the effect of the introduction of CO_2 into the plasma discharge on plasma energy, studies on the effect of CO_2 introduction on the plasma operational parameters, and characterization studies of the elemental detector for the packed column SFC.

SFC-HEMIP Interface. The interfacing of the SFC and the HEMIP involved placing the restrictor from the SFC into the plasma torch so that the SFC effluent could be swept into the plasma. The method of central tube introduction, described earlier, was the simpler method used, but proved to be the most troublesome. During the SFC separation, the restrictor would occasionally "freeze" because of the Joule-Thompson effect of the decompressing CO_2. This occurrence would cause a large "spike" in the detector output. A second area of concern was the possibility of the arcing of the plasma to the restrictor. The restrictor was intentionally placed near the plasma so that the tip could be warmed and the spiking effect reduced. When arcing occurred, the tip of the restrictor was fused shut. This problem was also noted by Hieftje et al. [4]. A third problem that was encountered with this arrangement was the instability of the plasma at low gas flows. Rates of 3 L/min were required to maintain a stable, well centered plasma.

A more satisfactory sample introduction method involved the use of a Pyrex "T" to mix the SFC effluent with the He plasma gas prior to introduction into the side-arm of the plasma torch. This arrangement (shown in Figure 3) allowed sufficient mixing of the two gases and permitted a stable plasma to be sustained with only 1 L/min plasma gas. The plasma gas flow, being at a right angle to the restrictor, was observed to prevent the nucleation of CO_2 as it decompressed and to eliminate spiking. This lower plasma gas flow rate resulted in less dilution and a longer residence time of the analyte in the plasma discharge. Also, arcing of the plasma to the restrictor was eliminated. These benefits indicated the sidearm arrangement as the appropriate interface to study for SFC-HEMIP.

The Effect of CO_2 on Plasma Energy. The He HEMIP has been characterized in terms of excitation temperature and electron number density in order to assess the effect of the addition of CO_2 on the energy of the plasma. The ionization temperature is also reported. Sample introduction methods included side-arm introduction for SFC and the direct introduction of aqueous samples with a pneumatic nebulizer/spray chamber system.

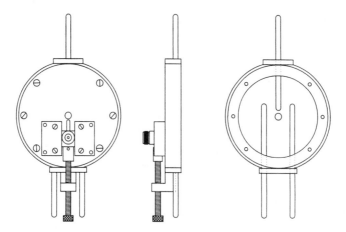

Figure 1:   Diagram of the HEMIP cavity showing top, side, and internal views.

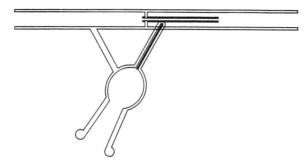

Figure 2:   Diagram of the tangential flow torch.

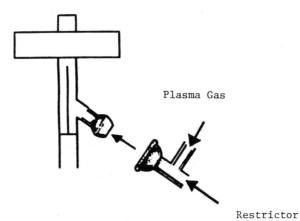

Figure 3:   Diagram of "T" connector for sidearm introduction of SFC effluent. (Adapted from ref. 19)

Excitation Temperature. The excitation temperatures were determined using several thermometric species (He with and without water and aqueous Fe). A listing of these calculated temperatures is presented in Table I. The temperature of the 150 W

Table I. Excitation Temperatures by Slope Method

Thermometric Species	Temperature (K)
He	6300
He + CO_2	6100
He + H_2O	5800
Fe (aqueous)	5600 (axial)
	5800 (radial)

Source: Adapted from refs. 12 and 17

He plasma was determined to be 6300 K from the radiance of He lines. The introduction of CO_2 at a flow rate of 120 mL/min (decompressed volume) yielded a temperature of 6100 K. The temperature obtained with direct introduction of water at 1 mL/min was determined to be 5800 K utilizing the the He lines and 5800 K (radial) and 5600 K (axial) from the Fe lines.

The values obtained from the He lines are not statistically different as one random error for the estimation of the excitation temperatures is 180 K. The effect of the introduction of CO_2 is more pronounced if Ar is used as the plasma gas. A 30% reduction in the temperature has been noted for a 150 W Ar HEMIP under similar conditions [5].

In comparison with other MIP systems, a temperature measurement for a Beenakker cavity sustaining a He plasma at 130 W of 4500 K has been reported [28], thus the excitation temperature of the He HEMIP is higher. This increased excitation temperature is an indication of the robustness of the He HEMIP and is also an indication that this cavity may be closer to sustaining an LTE discharge than any other other microwave cavity.

Electron Number Density. The electron number density was calculated for the HEMIP using the equations described by Griem [27]. These values for the number density are at best an approximation due to the resolution of the monochromator employed. They can be used to examine the effect of the introduction of CO_2 on the n_e- in the HEMIP, but should not be used in a rigorous comparison with electron densities of other plasma sources. The n_e- using the H_β emission line for a 150W He HEMIP was determined to be 5.3 X 10^{15} e-/cc. The introduction of 70 mL/min CO_2 or water at 1 mL/min from the nebulizer had no significant effect upon the calculated number density. However, if Ar is used as the plasma gas for the HEMIP at these same power levels, a significant lowering (factor of 10) in the number density value occurs [5].

Ionization Temperature. The ionization temperature was obtained for the He HEMIP by measuring the spatially integrated intensities of the cadmium emission atom (228.8 nm) and cadmium ion (226.5 nm). The samples were not introduced via the SFC system, but rather with a pneumatic nebulizer. Results of the ion/atom line ratios were tabulated and substituted into the Saha-Eggert relationship and T_{ion} as 6200 K calculated by an iterative process.

The Effect of CO_2 on Operational Parameters. The operational parameters examined with the introduction of CO_2 to a He HEMIP are the plasma geometry, the effect of pressure programming on the background signals, the effect of applied power on S/N, and the effect of plasma gas flow and chromatographic pressure on the S/N ratio.

The Effect of CO_2 on Plasma Geometry. Although the He HEMIP was shown to be unaffected in excitation temperature or electron number density by the addition of CO_2 into the plasma, the plasma discharge was observed to decrease in diameter. When operated solely with He or the addition of nebulized aqueous samples, the plasma discharge filled the 6 mm inner diameter of the plasma torch. The addition of CO_2 results in a discharge that was 3 mm in diameter. This reduction can be seen in Figure 4 where the S emission signal (axial view) from repetitive thiophene injections is monitored as the cavity is translated on its radial axis. The maximum emission occurs at the center of the discharge where the energy density is the greatest and symmetrically decreases to each side of the maximum. The signal could not be distinguished from the background beyond 1.5 mm from the center. The background was observed to be relatively constant across the width of the discharge.

The reduction in plasma diameter is attributed to the large percentage of CO_2 in the gas flow entering the discharge. For determinations using a packed column and a 1 L/min He gas flow, the CO_2 made up nearly 20% (by volume) of the gas.

Pressure Programming Effect on Background. Pressure programming is often used in SFC to aid in the separation process. Although this technique is of benefit to the separation, it can greatly complicate the spectrometric determination of the analyte species. The dissociation of CO_2 in the plasma produces several molecular products, which yield broad band emission. The effect of pressure programming on the background (P line at 255.3 nm) is shown in Figure 5. In plot A the background level is shown to increase linearly with the pressure until a value of 270 atm is reached. At this point a loss of coupling between the cavity and the generator occurs due to the increased mass of CO_2 being introduced into the plasma. This loss of coupling is noted by an increase in the reflected power from less than 1 W to 10 W.

To compensate for this effect, the cavity was adjusted to achieve critical coupling at 350 atm and the pressure program rerun, the cavity not being adjusted

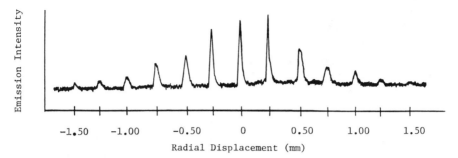

Figure 4: Plot of sulfur emission signal (581.9 nm) vs. axial viewing position for repetitive thiophene injections.

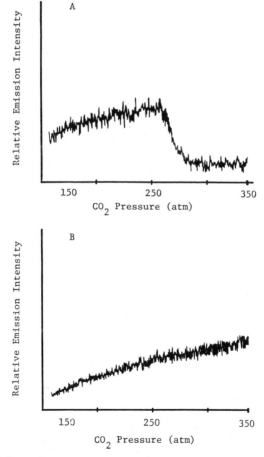

Figure 5: A, Effect of pressure programming on background emission for an improperly tuned cavity. B, Effect of pressure programming on background emission on a properly tuned cavity. (Adapted from ref. 12)

during the course of the run. Under these conditions the background level linearly increased with CO_2 pressure and the reflected power remained below 1W. Therefore, by properly tuning and matching the cavity for the upper limit of CO_2 flow obtained in the pressure ramp, the molecular background (which may comprise 10% of the signal) can be properly compensated.

Effect of Applied Power on S/N. The effect of applied power on the S/N ratio of the S emission (581.9 nm) for a thiophene sample is shown in Figure 6. Below 150 W, the plasma is not stable at the higher CO_2 pressures, but is stable at and above 150 W. However, at powers greater than 150 W the noise level increases and causes the S/N to degrade. The power level of 150W was thus chosen for this work, it being noted that the cavity and connectors did not become warm. At the higher applied levels the cavity was subject to heating and the reflected power was less than 1 W. The use of power levels greater than 150 W does not result in a more energetic plasma for the HEMIP cavity.

Effect of Plasma Flow and Chromatographic Pressure on S/N. In using packed columns for SFC separations, the CO_2 flow rate (decompressed) may approach a flow rate of 120 mL/min. This flow may constitute a significant portion of the plasma gas flow and may affect the S/N ratio. Illustrated in Figure 7 is a 3-D plot of He gas flow and SFC pressure on the S/N ratio of a Cl sample. For the data presented using He flow rates of 2 and 3 L/min, any difference in the S/N ratios is not statistically significant. At a He flow rate of 1 L/min a significant increase in S/N occurs at low CO_2 pressures. Under these conditions, the total gas flow (He and CO_2) is the least, thereby producing the least dilution of the analyte and its longest residence time in the discharge. The use of higher pressures results in a degradation of S/N. For pressure programming studies, a He flow rate of 2 L/min is used.

Characterization of the HEMIP. The HEMIP has been characterized in terms of linearity, and minimal detectable quantities for nonmetals and metals.

Linearity. The linearity ranges obtained using the HEMIP are from 3 to 5 orders of concentrative magnitude. Nonmetal values (P, Cl and S) range from 3 to 4 orders of magnitude. Metal determinations using the radial mode typically span 5 orders of concentrative magnitude. The upper limit of the working curve is limited by the loading of the column.

Reported MDQs. The MDQ values for S (thiophene), Cl (methylene chloride) and P (Paraoxon) are listed in Table II. These values are calculated by the method outlined by Hartmann for chromatographic detectors [24]. Under the experimental conditions for a packed column SFC determination, values of 50 ng for Cl, 30 ng for S, and 0.13 ng for P were determined while monitoring the lines of 479.5 nm, 581.9 nm, and 255.3 nm respectively. The deep UV line of S at 180.7 nm was also evaluated Additionally, the near IR lines for S were also studied, these have been recommended

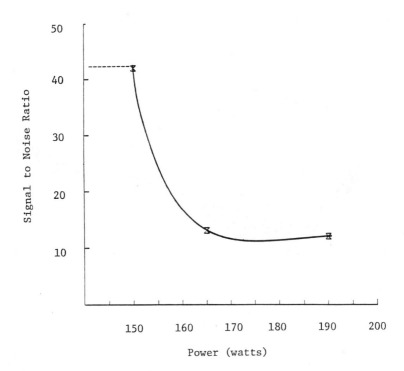

Figure 6: Effect of Applied Power on S/N ratio. Dotted line represents unstable plasma conditions. (Adapted from ref. 12)

as the better analyte emission lines for S due to the reduced background of the He plasma [4]. With the use of an IR grating and cutoff filters, an MDQ of 30 ng was obtained. These MDQs values for the different lines are not statistically different.

MDQ values for metals using an Ar HEMIP (150W, radial view) are also noted [19]. Fe was determined from ferrocene while Sn was determined from an industrial sample containing trichloromethyl tin.

It should be pointed out that limiting values for other element specific SFC detectors have been reported. A comparison of these values with those listed in Table II cannot readily be made due to the different manner in which many chromatographic limits of detection are calculated. Unlike c_L or x_L values, whose guidelines for calculation are carefully described for atomic spectrometric measurements [29], chromatographic limits of detection measurements are calculated by many methods and do not adhere to a central definition. This lack of definition is partially due to the differing behavior of many types of chromatographic detectors (mass vs. concentration dependent). Therefore, the values reported for packed column SFC conditions in Table II are not compared to other values such as listed in references 4 and 6.

Signal to Noise Ratio

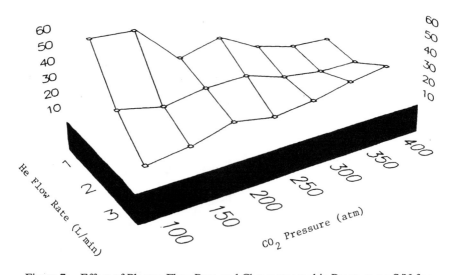

Figure 7:   Effect of Plasma Flow Rate and Chromatographic Pressure on S/N for Sulfur (581.9 nm). (Adapted from ref. 12)

Table II. MDQ for Packed Column SFC-He HEMIP

Element	Wavelength (nm)	MDQ (ng)
Cl	479.5	50
P	255.3	0.1
S	180.7	50
	581.9	30
	921.2	30
Fe	373.8	0.01[a]
Sn	270.78	0.03[a]

Source: adapted from ref. 12 and 19
[a]obtained using Ar

Conclusions

In this work, the HEMIP cavity, using He as the plasma gas, has been evaluated as an element selective detector for packed column SFC. A simple interface that permits a stable, low flow He plasma to operate has been developed. The high CO_2 flow rates experienced with packed column SFC do not affect the excitation temperature or the electron number density of the 150W HEMIP. Pressure programming of a packed column SFC can be conducted with the HEMIP with little effect on the S/N ratio and the cavity can be adjusted to achieve critical coupling over the entire pressure programming range. Although the plasma constricts with the addition of CO_2 to the plasma discharge, good MDQ values for Cl, P and S are reported. The use of UV or near IR lines for the nonmetals does not result in significantly lower MDQ values.

Acknowledgements

The authors would like to acknowledge the Virginia Water Resource Research Center, the Department of Interior-Bureau of Mines, and BP America Inc. for the financial support of this project. Also to be acknowledged are the Phillips Petroleum Company for the gift of the Holiday microwave generator, the Suprex Company for the loan of the Suprex SFC 200/A pump, and Keystone Scientific for the gift of the Deltabond columns. The authors also wish to thank Prof. Charles B. Boss of North Carolina State University, Prof. Larry T. Taylor of Virginia Polytechnic Institute and State University and Dr. Medhi Ashraf-Khorassani of the Suprex Corporation for helpful discussions, and Dr. Mark Wingerd and Keith McCleary for assistance with the computer generated figures.

References

1.   Olesik, J.; Olesik, S. *Anal. Chem.* **1987**, *59*, 796.
2.   Fujimoto, C.; Yoshida, H.; Jinno, K. *J. Chromatogr.* **1987**, *411*, 213.
3.   Skelton, R. J., Jr.; Farnsworth, P. B.; Markides; K. E. Lee; M. *Anal. Chem.* **1989**, *61*, 1815.
4.   Luffer, D.; Galante, L.; David, L.; Novonty, M.; Hieftje, G. *Anal. Chem.* **1988**, *60*, 1365.
5.   Motley, C. B.; Ashraf-Khorassani, M.; Long, G. L. *Appl. Spectrosc.* **1989**, *43*, 737.
6.   Carnahan, J. W.; Webster, G. K.; Zhang, L. 16th Annual Meeting of the Federal of Analytical Chemistry and Spectroscopist Societies, Chicago, IL; Oct. 1989; Abstra. No. 517.
7.   Jin, Q.; Hieftje, G.; Wang, F.; and Chambers, D. M. 1900 Winter Conference on Plasma Spectrochemistry St. Petersburg, FL; January 1990; Abstra. No. S-8.
8.   Bolton, J. S. Ph.D. Dissertation, Virginia Polytechnic Institute and State University, Blacksburg, VA, 1988.

9.   Blades, M. W.; Caughlin, B. L. *Spectrochim. Acta* **1985**, *40B,* 579.
10. Olesik, J. W.; Den, S. J.; Williamson, E. 13th Annual Meeting of the Federal of Analytical Chemistry and Spectroscopist Societies, St. Louis, MO, IL; Sept. 1986; Abstra. No. 90.
11. Perkins, L. D.; Long, G. L. *Appl. Spectrosc.* **1989**, *43,* 499.
12. Motley, C. B.; Long, G. L. *J. Anal. At. Spectrosc.* in press.
13. Michlewicz, K. W.; Carnahan, J. W. *Anal. Chem.* **1986**, *58,* 3122.
14. Wu, M.; Carnahan, J. W. *Appl. Spectrosc.* **1990**, *44,* 673.
15. Matus, L. G.; Boss, C. B.; Riddle, A. N. *Rev. Sci. Instrum.,* **1983**, *54,* 1667.
16. Burns, B.; Boss, C. B. *Appl. Spectrosc.* in press.
17. Long, G. L.; Perkins, L. D. *Appl. Spectrosc.* **1987**, *41,* 980.
18. Perkins, L. D; Long, G. L. *Appl. Spectrosc.* **1988**, *42,* 1285.
19. Motley, C. B.; Long, G. L. *Appl. Spectrosc.* **1990**, *44,* 667.
20. Smith, B. W.; Parsons, M. L. *J. Chem. Ed.* **1973**, *50,* 679.
21. Goode, S. R.; Kimbrough, L. K. *Applied Spectroscopy* **1988**, *42,* 1011.
22. Chester, T. L.; Innis, D. P.; Owens, G. D. *Anal. Chem.* **57**, 2243.
23. Deutsch, R. D.; Hieftje, G. M. *Appl. Spectrosc.* **1985**, *39,* 214.
24. Hartmann, C. H. *Anal. Chem.* **1971**, *43,* 113A.
25. Kalnickly, D. J.; Fassel, V. A.; Knisely, R. N. *Appl. Spectrosc.* **1977**, *31,* 137.
26. Blades, M. W.; Caughlin, B. L.; Walker, Z. H.; Burton, L. L. *Prog. Analyt. Spectrosc.* **1987**, *10,* 57.
27. Griem, H. R. "Spectral Line Broadening by Plasmas"; Academic Press: New York, 1974.
28. Abdallah, M. H.; Mermet, J. M. *Spectrochim. Acta* **1982**, *37B,* 391.
29. "Nomenclature, Symbols, Units, and their Usage in Spectrochemical Analysis II," *Spectrochim. Acta.* **1978**, *33B,* 242.

RECEIVED May 6, 1991

Chapter 15

Trace Selenium Speciation via High-Performance Liquid Chromatography with Ultraviolet and Direct-Current Plasma Emission Detection

William L. Childress[1], Donald Erickson[1], and Ira S. Krull[2]

[1]Winchester Engineering and Analytical Center, U.S. Food and Drug Administration, 109 Holton Street, Winchester, MA 01890
[2]Department of Chemistry and The Barnett Institute (341MU), Northeastern University, 360 Huntington Avenue, Boston, MA 02115

The U.S. Food & Drug Administration (FDA) is routinely called upon to monitor for levels of Selenium (Se) present in various commercial diet supplements, food and water supplies/sources. Most current methods involve determination of total Selenium, rather than individual species, but in recent years, it has become more apparent and required that regulatory agencies monitor and report individual metal species present.

This paper deals with the development of a new, fully validated trace method of analysis and speciation for inorganic compounds, an improved HPLC-UV/DCP approach providing information on volatile and nonvolatile Se species. This has involved direct HPLC-DCP interfacing, with introduction of ionic or non-ionic Selenium species into the DCP nebulizer and spray chamber. The reversed phase, paired-ion separations of ionic Se species could be followed, though not done here, by an on-line, hydride derivatization step, with introduction of volatile Se hydrides into the DCP.

Method validation has followed the determination of analytical figures of merit for the HPLC-Element Selective Detection (ESD) methods, using accepted practice, with standard addition or calibration plots for the quantitation of single blind, spiked and incurred samples. Several samples were possible for study, according to those most often regulated and monitored by FDA laboratories, such as dietary/food supplements, shellfish, plants, spices, and other foods or beverages. We were especially interested in determining the species and levels of Se compounds currently marketed in animal feed premixes.

0097–6156/92/0479–0257$06.00/0

The U.S. Food & Drug Administration (FDA) has, in part, a responsibility to continuously monitor foods and mineral supplements, eventually destined for human consumption. Unfortunately, there is as yet insufficient data to conclude what levels and how often Selenium (Se) species are to be found in our food supply. It was, in part, the purpose of this study to develop improved HPLC-element selective detection (ESD) approaches for Se species in man/animal dietary supplements, and to then utilize such newer, validated approaches for as many real world samples as possible and practical.

Se is an essential trace element in the human diet, and is often consumed as a dietary supplement in the form of Se tablets, mainly selenate (1-3). Some plants can absorb and store selenium from the soil in large amounts, and these may contain quantities from 1,000-10,000 ppm, dependent upon the plant. Se can complex with plasma proteins and can be distributed to all tissues of the body. Overexposure to selenium causes irritation of the eyes, nose, throat, and respiratory tract. It can, depending on the particular form or species present, cause cancer of the liver, pneumonia, degeneration of the liver and kidneys, and gastrointestinal disturbances. As such, it falls within the jurisdiction of the FDA to routinely monitor levels of selenate and/or other selenium species or derivatives that may be present in commercial formulations or the natural diet.

Current methods for the determination of Se species generally involve non-specific colorimetric tests (AOAC, USP) or total Se determinations by atomic absorption, flame or flameless (FAA/GFAA/AAS), atomic emission spectroscopy (AES), or electroanalytical techniques (4-6). A more acceptable method may be to generate the hydride and introduce this into a hydrogen-argon flame (FAA), together with an electrodeless discharge lamp and deuterium background correction (25). Se can also be determined using dialkyldithiocarbamate chelate formation and extraction steps from a food/beverage/drug sample, together with GFAA (1).

Most of the currently employed analytical protocols used within The Agency (FDA) are non-specific for different Se species, but measure only total Se. Thus, organoselenium or other oxidized forms of selenium (selenite/selenate) may go undetected or unspecified, unless some type of separation is combined with the final element selective detection step (AAS/AES). There are a number of viable and attractive metal speciation approaches that have already been described in the literature, specifically for Se species, but apparently none of these have ever been evaluated for possible applications to regulatory samples and requirements (7-10).

Within recent years, FDA/WEAC analytical laboratories have acquired extensive experience and expertise with regard to chromatography-direct current plasma (DCP) emission detection for metal species (11-18). Such approaches have involved both gas chromatography (GC) and high performance liquid chromatography (HPLC), often interfaced with metal specific DCP detection. These efforts have involved the development of an interface, optimization, evaluation, and final application of GC-DCP and/or HPLC-DCP for metals such as As, Sn, Cr, and Hg, but not Se.

In each case, we have demonstrated the unique capabilities, especially selectivity and sensitivity, that GC/HPLC-DCP interfacing can provide for trace metal determinations and speciation. The area of trace metal speciation by GC/HPLC-ESD has been reviewed on several occasions (19-24).

There exist different analytical approaches for Se speciation, though none of these uses HPLC-DCP. Total Se contents in various marine or body fluid samples have been determined, usually via hydride generation together with AAS techniques (25-27). This has led to significant reductions in minimum detection limits (MDLs) for most Se species capable of being converted to selenite, which is the only form which can be easily and quantitatively reduced to the hydride. Though these techniques are widely accepted and applied, they can only give total Se content, in the absence of any prior chromatographic separation. There are some electroanalytical techniques also reported for the determination of selenate or selenite in aqueous samples, but it is not yet clear that this approach can be used for other Se containing species (28).

Gas chromatography has been used infrequently to speciate for Se, and most often it has required the initial, usually off-line, pre-injection formation of a more volatile Se derivative. Typically, a complexation or chelation derivatization is performed in an aqueous sample to complex inorganic Se species, such as selenite, and the organic soluble derivatives are then extracted into an organic solvent, such as toluene (29). This Se-complexed solution is then directly injected into GC, usually with a non-element selective detector, such as ECD. Additional sample work-up reactions were necessary to then quantitate for other forms of Se. Clearly, this was a less than ideal GC Se speciation approach.

The most widely practiced Se speciation methods have been those involving HPLC together with some type of ESD. Most literature on Se speciation utilizes some type of HPLC or ion chromatography (IC) followed by either AAS or AES detection (30-33). In the absence of a hydride formation step, detection limits for most Se species using GC or HPLC directly coupled to either AAS or AES are not adequate for practical applications to most real samples. This is due to the somewhat limited volatility of most inorganic or organic Se species for GC or in the spray chamber or nebulizer of an ICP or DCP instrument. The increased volatility of a gaseous species, such as a metal hydride, appears to significantly improve the DCP response. This may be due to improved mass/hydride transfer over simple nebulization and aqueous vapor transfer. Just as direct AAS or AES detection limits can be significantly reduced by batch hydride formation, so too can speciation detection limits via HPLC-HY-AAS/AES methods. What is not yet clear from the literature, however, is the suitability of on-line hydride derivatization for Se species other than selenite.

Recently, we reported an HPLC-direct current plasma (DCP) emission spectroscopic approach for methyltins, using a novel type of paired-ion, reversed phase HPLC separation (14, 18). Because direct HPLC-DCP interfacing could not provide suitably low detection limits for some Sn species in certain samples, we and others have utilized a post-column hydride formation step after the separation and before introduction into the DCP plume (HPLC-HY-DCP)(14, 24). It seems likely that this same approach could/would prove suitable for Se species, if an HPLC-HY-DCP confirmatory method were needed or desired beyond an HPLC-DCP approach.

For naturally volatile organometals, or for those species that can readily be converted into volatile derivatives, GC still seems a most reasonable and practical approach. We described a GC-DCP approach for volatile organomercury species found in fish, especially for methylmercury (17). Thus, the combination of GC with DCP appears to remain a very reliable and practical approach to obtain "selective" chromatograms from injections of food supplement extracts. Most reports used some type of pre-injection derivatization for inorganic and other Se species. These have generally used hydridization, in order to provide improved GC performance characteristics of the original species. Improvements in off-column derivatization methods have recently used reaction GC to form hydrides of tributyltin and its analogs prior to flame photometric detection (6, 8, 34-35).

We report here on the direct interfacing of paired-ion, reversed phase HPLC methods for the initial separation of inorganic Se species, selenate and selenite, followed by an on-line, real-time, continuous interfacing with DCP. Direct interfacing is via a short, flexible Tefzel connector, which permits continuous HPLC effluent introduction into the DCP spray chamber at conventional flow rates for inorganic anion separations to be effective. The overall approach, HPLC-DCP, is very similar to what has been reported for Cr and Sn speciation (11, 14). Optimization of the interface has been followed by the derivation of analytical figures of merit for selenate and selenite, as well as by the demonstration of reproducibility of retention times, peak heights, and peak areas. Both on-line UV and DCP detections have been utilized, together or alone, for typical separations of mixtures of the two pure, inorganic Se species. Single blind spiked sample analyses of selenate/selenite in pure water have been conducted, in order to demonstrate quantitative capabilities prior to real world sample analyses of animal feed premixes. These real samples were then prepared by the FDA laboratory in Denver, spiking selenate and/or selenite at known levels. A comparison of total quantitative levels determined by HPLC-DCP and direct-DCP methods was made in order to further validate the HPLC-DCP approach. Because of the levels of Se containing species normally found in diet supplements for animals, above low ppm values, it has not been necessary to perform any post-column hydride formation prior to DCP detection.

Experimental

Chemicals, Reagents, and Solvents. Sodium selenite (IV) and sodium selenate (VI) (Alfa Ventron, Danvers, MA, USA), were dried overnight at 105'C and used without further purification. Stock solutions containing 1000 ppm Se were prepared in deionized/distilled water as needed, with working dilutions made daily. Methanol and acetonitrile were HPLC grade (J.T. Baker Chemical Co., Philipsburg, NJ). Water was

deionized and distilled in-house. HPLC mobile phases were filtered through Nylon-66 0.45 um membranes (Rainin Instrument Co., Woburn, MA, USA).
Ion-pairing reagents for the HPLC mobile phases were obtained commercially and used without further purification. Typical reagents used were Q5 (pentyltriethylammonium phosphate), Q6 (hexyltriethylammonium phosphate), Q7 (heptyltriethylammonium phosphate), Q8 (octyltriethylammonium phosphate), and Q12 (dodecyltriethylammonium phosphate) (Regis Chemical Co., Morton Grove, IL, USA). Other reagents included: tetrabutylammonium hydroxide (Fisher Scientific, Inc., Fair Lawn, NJ), tetrabutylammonium dihydrogen phosphate (Aldrich Chemical Co., Milwaukee, WI), tetrabutylammonium hydrogen sulfate (TBAHS, Fluka Chemical Co., Ronkonkoma, NY).

Instrumentation and Apparatus. HPLC systems were modular in design, consisting of a pump (Model M45 or M6000A, Waters Chromatography Division, Millipore Corp., Milford, MA, USA; Model 101A or 110A, Altex Division of Beckman Instruments Corp., Berkeley, CA, USA), an injector (Model 8780 Autosampler, Spectra-Physics Corp., San Jose, CA, USA; Model U6K, Waters; or Model 7125, Rheodyne Corp., Cotati, CA, USA), a UV detector (Model 757, Kratos Instruments, Division of Analytical Biosystems, Inc., Ramsey, NJ, USA), and an integrator (Model SP4270, Spectra-Physics or Model 3390A, Hewlett-Packard Corporation, Avondale, PA, USA).
Direct current plasma emission spectroscopy was carried out on a Spectraspan IIIb DCP unit operated in the diagnostic mode (Applied Research Labs, Valencia, CA, USA). The DCP was interfaced to a flowing stream from the HPLC system by connecting the column outlet tubing (Tefzel, 0.007 in i.d., Alltech/Applied Science) directly to the DCP spray chamber inlet via a short (about 1 cm) piece of the standard DCP spray chamber inlet tubing, a flexible Tygon tubing. The outlet tubing fit snugly into this short connector with no leaks observed with normal HPLC flow rates over prolonged periods of time, Figures 1-2.

HPLC Columns and Mobile Phase Conditions. HPLC columns were of the C-18 type, including: Nova-Pak C-18 (4um, 15cm x 3.9mm i.d., Waters), uBondapak C-18 (10um, 15cm x 3.9mm i.d. Waters), Ultrasphere C-18 (5um, 15cm x 4.6mm, Beckman), and Econosphere C-18 (5um, 15cm x 4.6mm i.d., Alltech/Applied Science, Inc., Deerfield Park, IL, USA). Columns were used at ambient temperature, without an external thermostat or controller. A short column (2cm x 3.9mm i.d., 30um) packed with LiChrosorb Si60 (EM Science, Inc., Gibbstown, NJ, USA) was installed between the HPLC pump and the injector. This saturated the mobile phase with silica and extended the lifetime of the analytical column when working at relatively high pH values (6-7.5) of the mobile phase. A guard column (Brownlee RP-300, C-8, 7um, 300A, 3cm x 4.6mm, Rainin Instruments) was placed ahead of the analytical column.

Animal Feed Premix Sample Preparation and Extraction. Animal feed premix samples were first sonicated (5 mins) in water or mobile phase, and then extracted by shaking at room temperature for about one hour. The suspension was centrifuged (10 mins at 2,000 rpm), filtered, and the supernatant withdrawn for HPLC injections. Aliquots, 50ul of this clear solution were injected for each HPLC analysis, following membrane

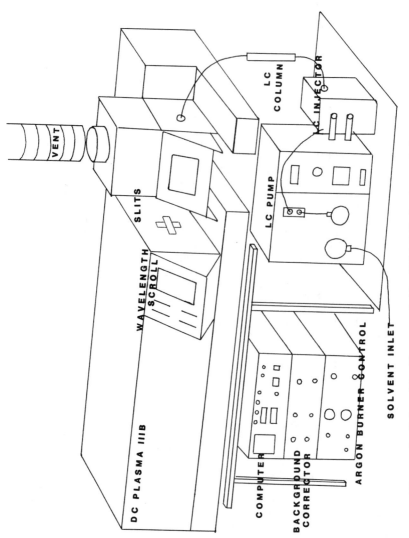

Figure 1. Schematic diagram of the HPLC–DCP instrumental arrangement, indicating placement of the HPLC eluent entry into the DCP.

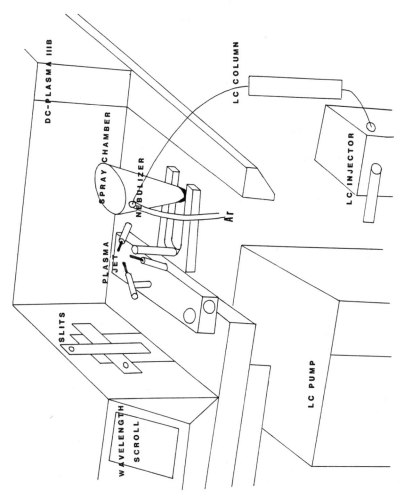

Figure 2. Schematic diagram of the spray chamber arrangement near the DCP plasma region, indicating the specific placement of the HPLC eluent entry into the spray chamber.

filtration (nylon-66, 0.45um, Genex, Inc., Gaithersburg, MD). A matrix matched calibration plot was derived by spiking a blank (no incurred Se) of the sample with known levels of each Se species (100, 200, and 300 ppm). These matrix-matched, spiked samples were treated in the same manner as the incurred samples. Incurred samples already contained some Se species, prior to spiking. Total Se content was determined using a batch-DCP method, following extraction and digestion procedures described elsewhere (39). It was necessary to prepare matrix-matched standards in order to perform these total Se quantitations, as done for the HPLC-DCP Se specification studies.

Results and Discussion

It was clear from previous results that the direct interfacing of HPLC separations with DCP detection would be immediately viable for Se species, after separation by conventional HPLC approaches, such as ion-pairing (11-16, 22, 36-38). We previously demonstrated that Cr (III) and Cr(VI) species could be easily separated under ion-pairing, RP-HPLC conditions, followed by a direct interfacing with DCP (11). In that study, two different ion-pairing conditions were needed, and the final detection limits were in the low ppm range. For methylated tin species, detection limits by direct HPLC-DCP were not adequate for real world samples, and thus an on-line, post-column, continuous hydride generation step was interfaced between the HPLC separation and DCP detection to provide HPLC-HY-DCP methods (14). In the case of the Se specification studies, real world samples of feed additive for animal supplements routinely contain one or both Se species at the low-mid ppm range (10-100 ppm or above). It was clear that the direct HPLC-DCP approach would prove adequate for such samples, and therefore all initial studies and optimizations did not involve any post-column hydride formation steps.

Optimization of HPLC Separation Conditions. The initial HPLC optimization experiments were performed using UV detection at 205nm, in order to determine the success of the HPLC separations for selenite (IV) and selenate (VI). Whereas selenate has a low absorptivity at 205nm, it was adequate to detect it at relatively high ppm concentrations, so that separations could be monitored. Selenite has a fairly high absorptivity, so that its detection presented no problem. The separation of selenite and selenate could be achieved using quaternary ammonium ion-pairing reagents under aqueous/organic RP conditions with a C-18 stationary phase. Concentrations of the ion-pairing reagent were ca. 1-5mM, with changes in concentration having the expected effect on retention times, viz., higher concentrations produced longer retention times. Mobile phases were buffered at pH 6-7 for optimum separations, and methanol or acetonitrile could be used in concentrations of at least 10% as the organic modifier. However, with organic contents in excess of 10%, there was a significant increase in the noise level on the DCP. Figure 3 illustrates a typical HPLC-UV chromatogram for standard selenite and selenate species, with conditions as indicated.

Figure 3. HPLC-UV chromatogram of separation of standard selenite and
    selenate species (20 ppm Se) under paired-ion, RP conditions:
    C-18 column (10um, 15cm X 3.9mm i.d.), RP-300 C-8 guard
    column (7um, 3cm x 4.6mm), pre-column Lichrosorb Si60 (30um,
    2cm x 3,9mm); mobile phase of 2.5 mM TBAHS, 0.01M each
    dipotassium hydrogen phosphate and potassium dihydrogen
    phosphate, pH = 6.55; flow rate 0.5ml/min; UV detection at
    205nm; injections of 50ul of 20ppm each Se species as
    standards. Peak indentities: 1 = selenite; 2 = selenate;
    3 = system.

Analytical Figures of Merit in HPLC-UV for Standard Selenite and Selenate. Linearity of calibration plots and minimum detection limits (MDLs) were determined by injecting progressively lower concentrations of each species until the detection limit was reached. This was defined as a signal-to-noise ratio of about 2/1, using the absolute background/baseline noise level as reference. In HPLC-UV, both selenite and selenate showed linear responses at concentrations up to about 50 ppm (50 ul injections). Detection limits were ca. 0.1 ppm (100 ppb) for selenate and 10 ppb for selenite. Calibration plots were linear over these concentration ranges, with coefficients of linearity of at least 0.999.

Though MDLs via HPLC-UV were more than adequate for the determination of both Se species in animal feed premixes, it was suspected that interferences might create insurmountable problems for UV detection alone, since especially at 205nm, many interferences from sample components would be expected, possibly to the point of obliterating the analyte peaks. Even with Se standards, considerable baseline disturbance was observed, due to effects of the components of the mobile phase that also respond at 205nm. It was therefore necessary and desirable to investigate DCP detection, concurrent with UV in order to again illustrate the major advantages of element selective detection in HPLC. At the primary and strongest emission wavelength for Se (196.026nm), there are no significant interfering elements.

Analytical Figures of Merit in FIA-DCP for Standard Selenite and Selenate. In order to optimize the DCP performance in the HPLC interfacing, batch studies were first performed, using Se standards. Detection limits via batch-DCP were found to be approximately 0.1 ppm, which agreed with the manufacturer's specifications for Se. This work was done in neat aqueous solutions. When solutions of the two Se species were prepared in various mobile phases for ion-pairing HPLC separations, extremely high background currents were seen on the DCP, due to the nature of the organic modifier present, with methanol showing more of an effect than acetonitrile. Since the desired HPLC separation could be realized with no organic modifier present, it was decided to use unmodified aqueous mobile phases for the FIA-and eventually HPLC-DCP phases.

In the flow injection analysis (FIA)-DCP mode, several parameters were varied in the HPLC separation conditions, without the column present, to determine their effect on the DCP response for Se. Linearity and MDLs were determined for both Se (IV) and Se(VI) in the FIA-DCP mode. Both species yielded linear responses up to 50ppm (50ul injections), and the MDLs were about 0.5ppm (500 ppb) for each species. To determine the optimum flow rate entering the DCP spray chamber, a solution of 10ppm Se (IV) was injected at flow rates varying from 0.2-2.0 ml/min. The mobile phase was 5mM Q5 (pentyltriethylammonium phosphate), with 0.01M each of potassium dihydrogen phosphate and dipotassium hydrogen phosphate. Injection volumes were 50ul. The maximum DCP response was realized at flow rates between 0.4-0.6 ml/min, ideal for the HPLC separations needed. At higher flow rates, DCP responses dropped off, with the response at 2.0 ml/min being about half that at 0.6 ml/min. Below 0.4 ml/min, the peak became very distorted. Thus, all subsequent experiments were performed at 0.5 ml/min.

Studies were then undertaken to evaluate the DCP response for Se species made up in ion-pairing medium, in order to determine if micellar formation of inorganic species led to DCP enhancements (11, 38). A series of mobile phases was prepared, each containing 0.01M mixed phosphate buffer and 5mM ion-pairing reagent. The ion-pairing reagents were the Q series, i.e., triethylammonium phosphates with 5-8 and 12-carbon chains as the fourth alkyl group completing the tetraalkylammonium cation. Again, 50ul of a 10ppm Se ((V) solution was injected with each mobile phase via FIA-DCP. There was essentially no difference in the DCP responses for Q5-Q8. When Q12 was explored, the DCP response actually decreased by about 10%. Since the desired separation could be achieved by using the shorter chain reagents, and since HPLC retention could be controlled by adjusting the concentration of the ion-pairing reagent and buffer, one still retained the ability to adjust HPLC parameters in order to realize ideal separations on the C-18 column used. It is perhaps significant to realize that ion-pairing reagents, as a function of their chain length and perhaps concentration, can adversely affect the DCP responses. MDLs remained constant or decreased throughout these studies, suggesting that micellar formation for Se species does not lead to improved analyte response via heightened mass transfer to the plume region of the DCP. This was contrary to expectations and earlier results for Cr species, again via HPLC-DCP (11).

The above optimized HPLC separation conditions were next interfaced with optimized DCP operating conditions, realized in the FIA-DCP studies. Figure 4 illustrates a typical HPLC-DCP chromatogram for the separation-detection of both Se species at the levels and conditions indicated. It was apparent that the desired separation and detection of only Se species could be accomplished.

Single Blind Spiking Determinations of Se (IV) and Se (VI) in Water. Table I summarizes both the HPLC-UV and HPLC-DCP results for the qualitative and quantitative determinations of Se (IV) and Se (VI) spiked into water as a single blind validation of the overall methodology. All of these analyses were performed reproducibly (n=4) on two separate days, and the data represents averages ± standard deviations for all of the data obtained on each sample. In general, all of the results were in agreement with the known, spiking levels, though at times, percent differences (found-actual) could approach 10%. The majority of the results had percent differences that were <5% between determined and spiked levels. Standard deviations (±SD) were less than ±1.0, which suggested that extremely good precision and reproducibility in all quantitative measurements were realized. Even though the minimum detection limits by HPLC-DCP were only about 1 ppm, accurate and precise quantitations have been possible in all of these spiked water samples, which ranged from 7.7-30.6 ppm Se. Thus, there was no apparent problem in immediate utilization of the newer HPLC-DCP methodology for real world samples. Because of current FDA interest, we selected animal feed premix type samples to further validate the HPLC-DCP method.

Quantitative Determinations of Se (IV) and Se (VI) in Animal Feed Premixes. These samples were prepared at the FDA laboratory in Denver, CO, known levels of both Se (IV) and Se (VI) species being added to authentic animal feed premixes. The levels spiked were in the range of what is normally used in animal feed premixes, prior to final

Figure 4. HPLC-DCP chromatogram for the separation-detection of
selenite and selenate species under paired-ion, RP
conditions: C-18 column (10um, 15cm x 3.9mm i.d.); mobile
phase of 2.5mM TBAHS, 0.01M each dipotassium hydrogen
phosphate and potassium dihydrogen phosphate, pH = 6.55;
flow rate 0.5ml/min; DCP detection at 196.026nm; injections
of 50ul of 20 ppm each Se species as standards. Peak
identities: 1 = selenite; 2 = selenate.

TABLE I. DETERMINATION OF SELENITE AND SELENATE IN SPIKED WATER SAMPLES BY HPLC-UV AND HPLC-DCP METHODS[a]

HPLC-UV Results

Sample	Se(IV)	Actual	% Difference	Se(VI)	Actual	% Difference
C	12.8±0.1	13.0	-1.5	--	--	--
D	21.4±0.8	22.0	-2.7	--	--	--
E	7.7+0.2	7.5	+2.7	32.5±0.2	31.5	+3.2
F	--	--	--	15.0±0.5	15.0	0.0
G	24.5±0.1	25.0	-2.0	10.7±0.2	11.0	-2.7
H	--	--	--	9.0±0.6	10.0	-10.0

HPLC-DCP Results

Sample	Se(IV)	Actual	% Difference	Se(VI)	Actual	% Difference
C	12.4±0.4	13.0	-.46	--	--	--
D	21.8±0.7	22.0	-0.9	--	--	--
E	7.7±0.4	7.5	+2.7	30.6±0.8	31.5	-2.9
F	--	--	--	16.3±0.7	15.0	+8.6
G	24.4±0.9	25.0	-2.4	11.6±0.3	11.0	+5.5
H	--	--	--	10.4±0.3	10.0	+4.0

---

a. Reported as average ppm Se ± standard deviation for n = 4, determinations performed on two separate, consecutive days.

addition to the actual animal feed. The Denver FDA laboratory did not perform any Se speciation studies. The WEAC laboratory worked up these samples, and quantitated using a matrix-matching technique with a blank feed premix sample spiked at known levels of Se species. Table II summarizes the data obtained for levels spiked and individual Se specification. We have indicated average ± standard deviation values for all determinations (n=3). In general, the levels of Se species determined via HPLC-DCP were in good agreement with the actual levels spiked. Table III summarizes data obtained for total Se content via direct-DCP methods, together with the summation of HPLC-DCP Se species, along with the total Se species spiked. Again, all of these numbers are in good agreement, especially the results of the HPLC-DCP and direct-DCP total Se values. There is a 5-10% disparity apparent between these values and the actual levels spiked.

It was not possible to perform any qualitative or quantitative determinations via HPLC-UV, because the presence of other UV absorbing species in the elution regions of the Se species prevented their detection, Figure 5. Without further sample work-up and preparation, it was not possible to utilize any form of UV detection with these particular HPLC conditions for either Se species in these samples. In contrast, the HPLC-DCP chromatograms, Figure 6, for these same animal feed premix samples clearly demonstrated the presence of one or both Se species, dependent upon the sample. There was little or no baseline disturbance via DCP detection, as opposed to UV, and the only species peaks observed were at the expected/correct retention times for Se (IV) and/or Se (VI). This is a very clear demonstration of how and when DCP detection provides adequate sensitivity, detection limits, and analyte selectivity as well as specificity when interfaced with chromatography.

The above results clearly demonstrate overall method validation for the direct HPLC-DCP interfacing method, and especially the opportunities provided. Unequivocal Se species identification has been possible, along with accurate, precise, and reproducible quantitative determinations in very complex sample matrices. Although the detection limits with direct HPLC-DCP interfacing are, at times, less than ideal, being only ca. 1 ppm for animal feed premixes, these are more than adequate and practical. There has not been a need for further pre-concentration of sample extracts, nor for the use of post-column, on-line, chemical generation of more volatile Se hydrides. Future work will, however, attempt to utilize solid phase, post-column hydride formation methods that will permit real-time, continuous, automated, and non-dilution approaches for improved detection limits of these and other Se containing species, inorganic or organometallic.

Acknowledgements

This work was performed at the Winchester (WEAC), MA District Office of the U.S. Food and Drug Administration (FDA). We are grateful to the FDA for the opportunity to perform this work and to report these results. Acknowledgement is made to M. Alpert (FDA/WEAC) who prepared blind, spiked samples for method validation purposes, and to individuals within the Denver, CO FDA labs. Specifically, we wish to acknowledge the interest and capable assistance of J. Hurlbut (Science Advisor to DEN-DO), Metropolitan State College, Denver, CO, and C.

TABLE II. HPLC-DCP DETERMINATIONS OF SELENATE AND SELENITE IN SINGLE
BLIND SPIKED ANIMAL FEED PREMIX[a]

Sample	Se(IV)[b]	Actual	% Diff	Se(VI)[b]	Actual	% Diff
1	200.8±9.0	200	+0.4	---	---	---
2	---	---	---	207.2±5.0	200	+3.6
3	105.5±6.5	100	+6.5	113.7+3.2	100	+13.7

a. Quantitations done by matrix matched calibration plots for both Se
species, using blank sample containing no spiked/incurred Se
content.
b. Average ± standard deviation, n=3, given as ppm concentrations in
feed premix samples.

TABLE III. TOTAL SELENIUM CONTENT BY DIRECT-DCP IN SINGLE BLIND
SPIKED ANIMAL FEED PREMIX SAMPLES[a]

Sample No.	Total Se by Direct -DCP[b]	Total Se by HPLC-DCP[b]	Spiked
1	210.8±7.8	200.9±9.0	200
2	212.6±11.2	207.2±3.6	200
3	215.7±3.7	219.3±8.6	200

a. Quantitations done by matrix matched calibration plots for total Se,
using blank sample containing no spiked/incurred Se content.
b. Average ± standard deviation, n=3, given as ppm concentrations in
feed premix samples.

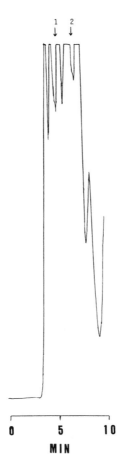

Figure 5. HPLC-UV chromatogram of feed premix sample containing
nominally 100ppm Se each as selenite and selenate species,
HPLC-UV conditions as in Figure 3.

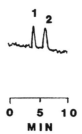

Figure 6. HPLC-DCP chromatogram of feed premix sample containing
nominally 100ppm Se each as selenite and selenate species,
HPLC-DCP conditions as in Figure 4. Peak indentities:
1 = selenite; 2 = selenate.

Geisler, supervisory chemist, DEN-DO of FDA. M. Finkelson, W.S. Adams, A. Falco, and others within FDA/WEAC also provided encouragement, time, and guidance during these studies. I.S. Krull is a current Science Advisor to the Winchester Engineering & Analytical Center District Office of the U.S. FDA.

This is contribution number 429 from The Barnett Institute at Northeastern University.

Literature Cited

1. Berman,E.. Toxic Metals and Their Analysis; Heyden & Son, Ltd.: London, England, 1980; Chap. 23, p.183.
2. Handbook of Carcinogens and Other Hazardous Substances; Bowman.M.C., Ed.; Marcel Dekker, Inc.: New York, 1982; pp. 656-658.
3. Casarett and Doull's Toxicology, The Basic Science of Poisons; Doull, J.; Klaasen, C.D.; Amdur, M.O., Eds.; Macmillan Publishing Co.: New York, 1980; Chap. 409.
4. Pinta, M.; Modern Methods for Trace Element Analysis; Ann Arbor Science: Ann Arbor, MI, 1978.
5. Physical Methods in Modern Chemical Analysis; Kuwana, T., Ed.; Academic Press, Inc.: New York , 1978, 1980; vols. 1 and 2.
6. Ultrace Metal Analysis in Biological Science and the Environment; Risby, T.H., Ed.; Advances in Chemistry Series No. 172; American Chemical Society: Washington, DC, 1979.
7. Liquid Chromatography in Environment Analysis; Lawrence, J.F., Ed.; The Humana Press: Clifton, NJ, 1984, FRG, 1981.
8. Schwedt, G.; Chromatographic Methods In Organic Analysis; A.H. Verlag: Heidelberg, FRG, 1981.
9. Florence, T.M.; Batley, G.E. In Critical Reviews in Analytical Chemistry; CRC Press, Inc.: Boca Raton, FL, 1980; p.219.
10. The Importance of Chemical Separation in Environmental Processes; Bernhard, M.; Brinckman, F.E.; Sadler, P.J., Eds.; Dahlem Conference (1984): Springer-Verlag, Berlin, FRG, 1986.
11. Krull, I.S.; Panaro, K.; Gersham, L.L. J. Chrom. Sci. 1983, vol. 21, p. 460.
12. Bushee, D.S.; Krull, I.S.; Demeko, P.R.; Smith, S.B., Jr. J. Liquid Chrom. 1984, vol. 7(5), p. 8671.
13. Panaro, K.; Krull, I.S. Anal. Letters. 1984, vol. 17(A2), p. 157.
14. Krull, I.S.; Panaro, K.W. Appl. Spec. 1985, vol.39, p.183.
15. Krull, I.S.; Bushee, D.; Schleicher, R.G.; Smith, S.B. Jr. The Analyst. 1986, vol. 111, p. 345.
16. Bushee, D.S.; Krull, I.S.; Smith, S.B., Jr.; Schleicher, R.G. Anal. Chim. Acta. 1987, vol. 194, p. 235.
17. Panaro, K.W.; Erickson, D.W.; Krull, I.S. The Analyst. 1987, vol.112, p.1097.
18. Panaro, K.W.; Erickson, D.; Krull, I.S. Appl. Organomet. Chem. 1989, vol.3, p.295.
19. Krull, I.S.; Jordan, S. American Laboratory. October 1980, vol.
20. Krull, I.S. Trends in Analytical Chemistry (TrAC). 1984, vol. 3(3), p.76.
21. Liquid Chromatography in Environmental Analysis; Lawrence, J.F., Ed; The Humana Press: Clifton, NJ, 1984; chap. 5.

22. Advances in Chromatography and Separation Chemistry; Ahuja, S., Ed.; ACS Symposium Series; American chemical Society: Washington, D.C., 1986; chap.11.
23. The Importance of Chemical Speciation in Environmental Processes; Bernhard, M.; Brinckman, F.E.; Sadler, P.J., Eds.; West Berlin, FRG, September 1984; Springer-Verlag, Berlin and Heidelberg, December, 1986; p.579.
24. Trace Metal Analysis and Speciation; Krull, I.S., Ed.; Elsevier Science Publishers: B.V., Amsterdam, The Netherlands, 1990-91.
25. Welz, B.; Melchner, M. Anal. Chem. 1985, vol. 57, p. 427.
26. Welz, B.J. Instrumentation - Research. 1986, vol. 46.
27. Willie, S.N.; Sturgeon, R.E.; Berman, S.S. Anal. Chem. 1986, vol. 58, p.1140.
28. Brimmer, S.P.; Fawcett, W.R.; Kulhavy, K.A. Anal. Chem. 1987, vol. 59, p. 1470.
29. Measures, C.I.; Burton, J.D. Anal. Chim. Acta. 1980, vol. 120, p.177.
30. Irgolic, K.J.; Stockton, R.A.; Chakraborti, D.; Beyer, W. Spec. Acta. 1983, vol. 38B(1/2), P.437.
31. Roden, D.R.; Tallman, D.E. Anal. Chem. 1982, vol. 54, p. 307.
32. Hillman, D.C. M.S. Thesis. Texas A&M University. August, 1981.
33. Shibata, Y.; Morita, M.; Fuwa, K. Anal. Chem. 1984, vol. 56, p. 1527.
34. Sullivan, J.J.; Torkildson, J.P.; Wekell, M.W.; Hollingsworth, A.; Saxton, W.L.; Miller, G.A.; Panaro, K.W.; Uhler, A.D. Anal. Chem. 1988, vol. 60, p. 626.
35. Krull, I.S.; Colgan, S.; Xie, K-H; Nueu, U.; King, R.; Bidlingmeyer, B. J. Liquid Chrom. 1983, vol. 6(6), p. 1015.
36. Dou, L.; Krull, I.S. J. Chromatogr. 1990, vol. 499, p.685.
37. Lookabaugh, M.; Krull, I.S. J. Chromatogr. 1988, vol. 452, p. 295.
38. Kirkman, C.M.; Zu-ben, C.; Uden, P.C.; Stratton, W.J.; Henderson, D.E. J. Chromatogr. 1984, vol. 317, p.569.
39. Burkepile, R.G.; Hurlbut, J.A.; Geisler, C.A. Laoboratory Bulletin No. 3144. Denver District Laboratory. U.S. Food & Drug Administration (FDA). May 13, 1987.

RECEIVED April 12, 1991

Chapter 16

Analytical Utility of an Inductively Coupled Plasma–Ion Chromatographic System for the Speciation and Detection of Transition Metals

Daniel J. Gerth[1] and Peter N. Keliher[2]

Department of Chemistry, Villanova University, Villanova, PA 19085

An overview of the construction and application of a coupled ion chromatograph-inductively coupled plasma instrument system is presented. The instrumentation is discussed in terms of equipment and software requirements, and species specific detection in Fe (II)/Fe (II) and Cr (III)/Cr (VI) systems is demonstrated.

The advantages of atomic spectrometric detection for chromatography have been discussed many times. Recent publications (1,2,3) indicate that the technique is rapidly gaining acceptance as an important tool for both speciation and discrete sample introduction.

In order to examine the potential of the technique, an ion chromatography/inductively coupled plasma instrument system (hereafter called IC-ICP), has been developed. A major objective of the project was to interface the instruments in such a fashion as to allow quick and simple changeover from the IC-ICP operating mode to continuous nebulization.

Equipment

The ICP utilized in this study was an Applied Research Laboratories QA-13700 Quantometric Analyzer (Sunland, CA), equipped with a 24 channel polychromator, and 2 Kw RF generator operating at 27.4 MHz. Standard operating conditions are given in Table I.

Table I. ICP Standard Operating Conditions

Forward Power	1.1 kw
Plasma Gas Flow	1.2 L/min
Coolant Gas Flow	10.0 L/min
Carrier Gas Flow	1.0 L/min

[1]Current address: Lancaster Laboratories, 2425 New Holland Pike, Lancaster, PA 17601–5994
[2]Deceased

0097–6156/92/0479–0275$06.00/0
© 1992 American Chemical Society

The instrument electronics have been upgraded by LABCO, Inc.(Califon, NJ) and consist of three eight channel 16 bit A/D boards, stepper motor driver, autosampler controller, and power supply. The instrument is controlled by an IBM-PC/AT compatible computer, with 1 megabyte RAM, EGA graphics, and a 20 megabyte hard disk drive. Communications are through a standard RS-232 port. All math is performed with double precision floating point numbers, which provide 15 to 16 significant digits.

The chromatographic equipment consists of a Beckman/Altex 110A HPLC pump, Rheodyne 9125 metal-free injection valve, and CS-5 separator column (Dionex, Sunnyvale CA). Various sample loop volumes were used during the course of the study. In keeping with our objective of minimal hardware modifications, the separator column was connected directly to a standard Meinhard concentric nebulizer mounted in a stock ARL conical spray chamber.

Software

The controlling software was written in Pascal (Borland International, Scotts Valley CA), which was chosen because of its flexibility, speed, and modular programming approach (4). The entire package consists of a small "driver" program, which initializes the system, and several self-contained modules, each containing all of the functions and procedures related to a specific task (i.e. calibration, analysis). Global variables were kept to a minimum. This approach has two main advantages : 1. Since the modules are used as overlays, the amount of program code resident in memory at any particular time is small, leaving more room for the manipulation of large data structures, and, 2. The modular approach allows new routines, such as the IC-ICP code, to be easily added or modified at any time.

The IC-ICP module consists of two procedures, Acquire and View. Acquire controls the actual data acquisition, allowing the operator to specify the integration time, run length, data storage file, and select a channel for real-time monitoring of the run. View allows the operator to display chromatograms from up to four channels simultaneously. Since the instrument is equipped with a polychromator, the number of channels actually stored depends upon the analytical program in operation at the time; this must be selected from the main menu before any other tasks may be performed. Up to the full compliment of 24 channels may be included.

For normal chromatographic runs, an integration time of 0.5 seconds provides adequate sampling, and is the default value. The actual time between data points is 1.4 seconds, due to the 0.9 seconds the instrument requires to read, transmit, and process the 24 channels in the polychromator array. In order to minimize this delay, we opted for a polling routine during the read cycle. As the instrument is integrating, the software stores the previously acquired points and updates the screen display. When finished, the software then continuously polls the serial port until it receives a data ready signal, at which point it accepts the new data and repeats the cycle.

Discrete Sample Introduction

Discrete sample introduction is defined here as the injection of small, fixed volume aliquots of sample into an instrument. The opportunity to obtain a complete ICP analysis on as little as 20 μL of sample is to us as compelling a reason for examination of IC-ICP as speciation. To a certain extent, the concentration detection limit of the IC-ICP system is adjustable by simply utilizing different sample loop volumes. The signal detection limit is determined by the baseline noise, which was measured by simply letting the instrument run without injecting a sample, and calculating three times the standard deviation of the baseline. The concentration detection limit was then obtained by relating the signal detection limit to intensities obtained by specifying a 1.0 mL sample loop, and injecting a mixed standard. Table II presents the signal and concentration detection limits for Fe (III) and Cr (III) obtained in this study, and the corresponding continuous nebulization detection limit.

Table II. Detection Limits

Analyte	3s Background Signal (counts)	1.0 mL Injection (mg/L)	Continuous Nebulization (mg/L)
Fe (III)	33.3	0.14	0.005
Cr (III)	26.1	0.090	0.007

As mentioned above, the technique's concentration detection limit may be controlled by varying the sample loop volume. Figure 1 (a) and (b) both were obtained by injections of the same 20 mg/L Cr (III) standard, using a 20 μL and 1.0 mL sample loop, respectively. The peak obtained with the 20 μL injection is easily seen, and while the peak shape has degraded somewhat, quantitation may still be performed via either peak height or area measurement. Since the ICP acts as a mass sensitive detector, theoretically a fivefold increase in the mass of analyte injected should result in a fivefold increase in the detector signal. As Table III illustrates, peak area exhibits a 5% deviation from the theoretical response, and is the measurement of choice.

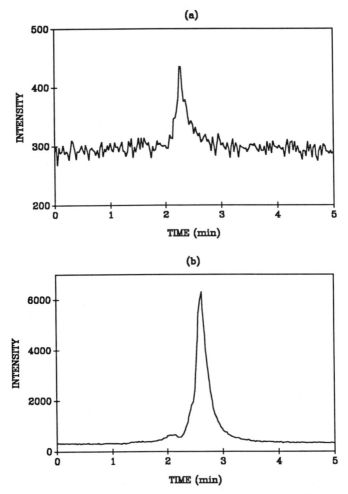

FIGURE 1. Sensitivity modification by changing sample injection loop size. Both chromatograms are of a 20 mg\L Cr (III) standard. (a) 20 μL sample loop, (b) 1000 μL sample loop.

Table III. A Comparison of Peak Height and Peak Area Quantitation

	20 μL Sample Volume	1000 μL Sample Volume	% Deviation from Theoretical
Peak Area (count*min)	1242	65656	5.0
Peak Height (counts)	140	6005	15.0

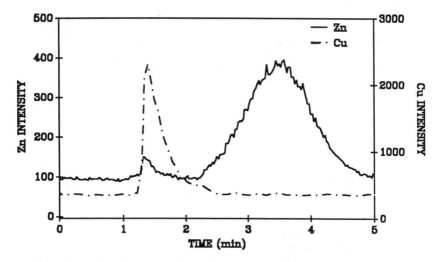

FIGURE 2. Simultaneous detection of 20 mg/L copper and
0.2 mg/L zinc. The small peak in the zinc trace is due
to spectral overlap of the copper 213.598 nm line.

Figure 2 illustrates another important aspect of the
technique. This chromatogram was obtained by injecting 100
μL of a mixed standard containing 20 mg/L copper and 0.2
mg/L of zinc, and simultaneously monitoring the Cu 324.754
nm and Zn 213.856 lines. The small peak in the zinc trace
co-eluting with the copper is due to spectral overlap of
the Cu 213.598 nm line, while the actual zinc peak is
completely resolved from the copper interference. Because
of this separation of analyte bands in time, most spectral
overlap problems are resolved, which greatly simplifies
analysis since monitoring all potentially interfering lines
is not necessary.

Speciation Studies

Due to the similarities in size and enthalpies of
hydration, most transition metals exhibit very similar
affinities for cation exchange resin (5). In order to
effect a useful separation, chelating agents are employed
as eluents. By varying the chelant and eluent pH, the
retention times of the various metals can be adjusted as
required. Some chelating agents commonly employed are
oxalic, citric, and tartaric acids, EDTA, and, more
recently, pyridine-dicarboxylic acid, or PDCA. Lithium

hydroxide is normally employed to adjust the eluent pH, because of the low affinity the Li^+ ion has for the ion exchange resin.

Iron. Iron speciation was carried out on a CS-2 cation separator column (Dionex, Sunnyvale, CA) using a 10 mM oxalic/7.5 mM citric acid eluent, adjusted to pH 4.3 with LiOH. Under these conditions, Fe (III) forms a triply charged anionic complex, and is unretained by the resin. Fe (II), however, is not strongly complexed, and elutes much later, at 11.5 minutes. Figure 3 demonstrates this separation.

Chromium. The importance of species specific information in environmental Chromium analyses cannot be understated. Since the typical chrome waste treatment employs a reduction step to convert the carcinogenic Cr (VI) to the more benign Cr (III), it is imperative to know all species concentrations in order to determine the efficacy of the treatment program. Direct chromatography of the two forms is impossible, however, due to the presence of various Cr (III) - hydroxyl complexes. Due to the slightly different affinity of each complex for the ion exchange resin, these complexes elute as a series of low, broad unresolved bands under the chromatographic conditions used in this study. Figure 4 is the chromatogram resulting from the injection of a solution containing 20 mg/L of Cr (III) utilzing a PDCA based eluent. As can be seen, it is impossible to quantitate the chromium.
It is known, however, that Cr (III) forms a stable, mononegative complex with PDCA near neutral pH (6). While this reaction is slow at room temperature, heating the sample to boiling for a brief period (under one minute) allows the reaction to proceed to completion. The resulting complex can then be easily separated from the Cr (VI), which, at the same pH, is predominantly present as the dinegative chromate anion.
In order to effect the complexation, we used a microwave digestion oven, and determined the optimum heating time at 100% power. All complexation reactions were carried out in open polyethylene Erlenmeyer flasks to avoid overheating and pressurization. Figure 5 shows the chromatograms after 10, 20, 30, 40, 50, and 60 seconds of microwave exposure. From a graph of peak height vs. time, 30 seconds exposure was chosen as optimum. The first peak (seen most clearly in 5a) is the monopositive Cr-PDCA cation.
Since oxidation of excess PDCA by chromate would lead to high Cr (III) results, we subjected a 20 mg/L solution of chromate to the same complexation procedure. Figure 6 clearly demonstrates that no oxidation takes place. Figures 7 and 8 are chromatograms of mixtures of Cr (III) and Cr (VI), demonstrating the final separation.
Future work planned includes examining industrial

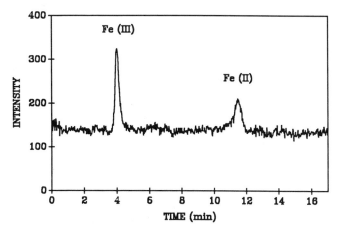

FIGURE 3. Separation of Fe (II) and Fe (III) utilizing an oxalate based eluent.

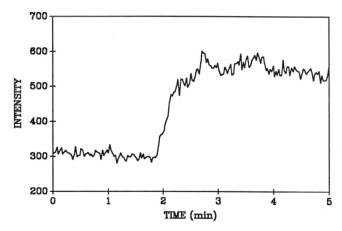

FIGURE 4. Chromatogram obtained by injecting 1000 μL of a 20 mg/L Cr (III) standard without prior complexation.

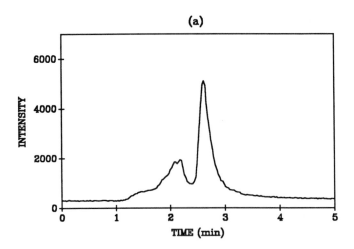

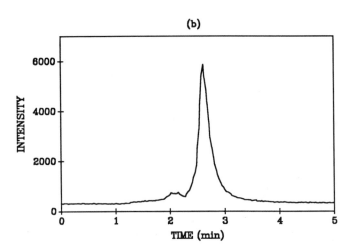

FIGURE 5.   Optimization of microwave heating time for
the formation of the Cr (III)-PDCA complex.   (a) 10
seconds, (b) 20 seconds, (c) 30 seconds, (d) 40
seconds, (e) 50 seconds, (f) 60 seconds.

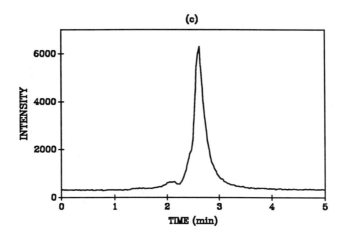

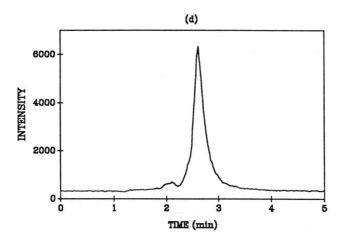

Figure 5. Continued. *Continued on next page.*

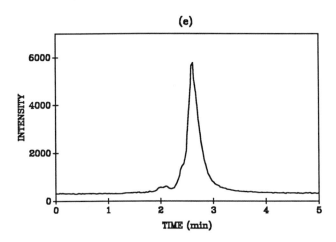

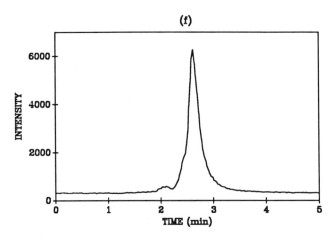

Figure 5. Continued.

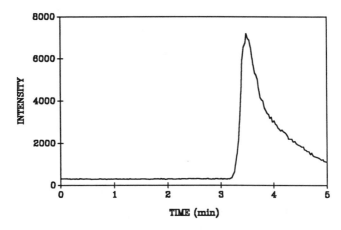

FIGURE 6.   A 20 mg/L Cr (VI) standard after undergoing the PDCA complexation procedure. The lack of a Cr (III) peak indicates that no oxidation of the PDCA by the Cr (VI) is taking place.

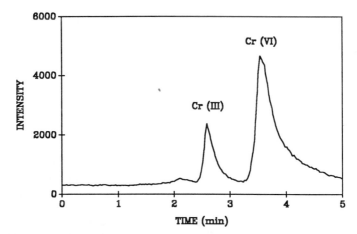

FIGURE 7.   Chromatogram obtained from a 1000 μL injection of a 10 mg/L Cr (III) - 10 mg/L Cr (VI) mixed standard. Cr (III) is eluted as the Cr(PDCA) anionic complex, Cr (VI) as the chromate anion.

(a)

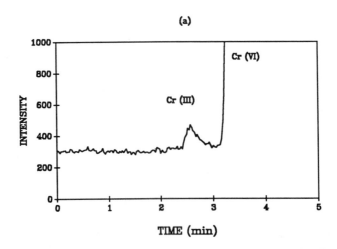

(b)

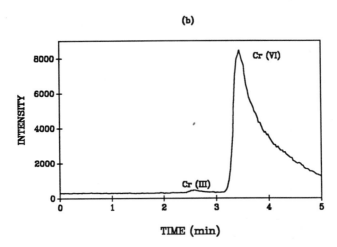

FIGURE 8.   Chromatogram obtained from a 1000 μL injection of a 0.2 mg/L Cr (III) - 20 mg/L Cr (VI) mixed standard. (a) scale expansion showing complete resolution of the Cr (III) peak. (b) both peaks on scale, illustrating the dynamic range of the technique.

boiler and cooling tower waters, a comparison of detection limits obtained with alternate nebulization systems (e.g. Hildebrand Grid nebulizer), and further augmentation of the software to allow complete post run processing of the chromatogram.

LITERATURE CITED

1.) Keliher, P.N.; Gerth, D.J.; Snyder, J.L.; Wang, H.; Zhu, S.F. Anal Chem, **1988**, 60, 342R-368R.
2.) Keliher, P.N.; Ibrahim, H.; Gerth, D.J. Anal Chem, **1990**, 62, 184R-212R.
3.) Environmental Analysis using Chromatography Interfaced with Atomic Spectroscopy; Harrison, R.M.; Rapsomanikis, S., Eds., Ellis Horwood Ltd.: Chichester, U.K., **1989**.
4.) Turbo Pascal Reference Manual, Borland International, Scotts Valley CA.
5.) Smith, R. E. Ion Chromatography Applications; CRC Press: Boca Raton FL, **1988**; Vol. 1, pg 77.
6.) Ion Chromatography Application Note 26; Dionex Corporation, Sunnyvale, CA, **1986**; pp 1-2.

RECEIVED April 26, 1991

Chapter 17

Chromatographic Detection by Plasma Mass Spectrometry

Lisa K. Olson[1], Douglas T. Heitkemper[2], and Joseph A. Caruso[1]

[1]Department of Chemistry, University of Cincinnati, Cincinnati, OH 45221
[2]National Forensic Chemistry Center, U.S. Food and Drug Administration, Cincinnati, OH 45202

Plasma source mass spectrometry, when coupled with an appropriate chromatographic technique, provides a method for trace elemental analysis with species selectivity. A review of investigations concerning chromatographic detection by plasma mass spectrometry techniques is presented. Both gas and liquid chromatographic methods combined with ICP-MS and MIP-MS detectors are discussed. For example, the determination of organotins and halogenated compounds with GC-MIP-MS results in low to sub-picogram detection limits. Phosphorus and sulfur containing compounds separated by GC have been determined with both He and N_2 MIP-MS systems with low nanogram to picogram detection limits. Additionally, the use of HPLC-ICP-MS to eliminate the $ArCl^+$ interference on the determination of arsenic is described. Finally, the detection of halogenated compounds with HPLC-MIP-MS is presented.

There is growing concern over the introduction of toxic trace elements into the environment. Their presence, even at trace levels, can adversely affect both the natural environment and the industrial workplace. In addition, the particular chemical form; organic, organometallic, or inorganic, is important in assessing the toxicity of the element. Therefore, it is increasingly necessary to establish species selective methods for their determination.

Plasma source mass spectrometry provides a means to analyze samples for various elements at sub-nanogram to picogram levels by combining the ability of a plasma to efficiently ionize elements with the sensitivity of a quadrupole mass spectrometer. In addition to the low levels of detection possible for many elements,

0097–6156/92/0479–0288$06.25/0

plasma mass spectrometry also has the advantages of multielement detection and the ability to obtain isotope ratio information.

Extensive use has been made of gas and high performance liquid chromatography for speciation of organic, organometallic and inorganic compounds. Plasma source mass spectrometry when coupled with an appropriate chromatographic technique provides a method for trace elemental analysis with species selectivity.

The subject of this chapter will be chromatographic detection by plasma mass spectrometry. A brief review of work done in this area by other researchers will be presented. The use of helium microwave-induced plasma mass spectrometry (MIP-MS) for detection of both metallic and non-metallic elements separated by gas chromatography (GC) will be presented. In addition, the use of both inductively-coupled plasma mass spectrometry (ICP-MS) and MIP-MS for detection of high performance liquid chromatography (HPLC) effluents will be discussed. Representative separations involving both metals and non-metals demonstrate the potential of both ICP-MS and MIP-MS to provide extremely low levels of detection with speciation information. In addition, chromatography is also shown to be useful in eliminating interferences that hinder determinations by plasma MS.

Plasma Mass Spectrometry

Plasma source techniques are widely used for trace element determinations. ICP-MS offers several unique advantages to atomic spectrometry. In addition to low levels of detection obtainable for most of the elements, ICP-MS is a multielement technique and although sequential, is extremely fast due to the rapid scanning capability of a quadrupole mass analyzer. These attributes, along with spectral simplicity and the ability to measure isotopic abundance account for the rapid evolution and popularity of ICP-MS. Those readers interested in an in-depth discussion of ICP-MS are directed to the many reviews available (1-7).

Recently interest in alternative plasma sources, such as the He MIP, has increased. In the early 1980's, Douglas and co-workers (8,9) described an Ar MIP-MS system which exhibited good results. However, there are advantages to using helium as the plasma gas. The argon plasma is not an efficient ionization source for the higher ionization potential elements such as the halogens, phosphorus and sulfur (10,11). Helium has a much higher ionization potential than argon (24.5 eV vs. 15.8 eV) and thus is a more efficient ionization source for many non-metals, thereby resulting in improved sensitivity.

Another advantage of using a helium plasma is the elimination of isobaric interferences arising from argon containing polyatomic species (12). The use of helium as the plasma gas should help eliminate these interferences by shifting them lower in the mass range since helium is virtually monoisotopic with m/z of 4.

The instrumentation and operating parameters in He MIP-MS vary to a greater extent from system to system as opposed to Ar ICP-MS instruments. No commercial MIP-MS instruments are currently on the market; thus using MIP-MS requires modifications to existing mass spectrometer systems. The actual operating conditions, including power and gas flow rates, depend on the type of source, i.e. atmospheric-pressure or low-pressure, and the type of sample introduction, i.e. gaseous or aqueous.

Plasma MS Detection for Gas Chromatography

In general, several features are necessary for a detector to be suitable for GC applications including adequate sensitivity, selectivity, linearity, rapid response and low dead volume (13). Plasma emission detectors have successfully been used as GC detectors providing sensitive element selective detection. A GC-MIP-AES system is currently on the market (14). Both ICPs (15-21) and He MIPs (13,15,22-26) have been utilized as emission detectors for GC. However, the increased sensitivity and isotopic information available from mass spectrometry makes the coupling of GC with plasma MS techniques a logical development. Only recently has interest grown in investigating GC-plasma MS.

GC-ICP-MS. Limited studies have focused on coupling GC with ICP-MS detection. GC is most commonly used to separate compounds containing non-metallic elements which are difficult to detect at the desired trace levels with ICP-MS because of their low degree of ionization in an Ar plasma. Peak broadening from diffusion in the torch can cause problems when interfacing GC to ICP-MS.

In one publication, Chong and Houk (27) described the coupling of packed-column GC with Ar ICP-MS to provide potentially sensitive element-selective detection as well as atomic ratios and empirical formulas. The information available could complement molecular information obtained with conventional GC-MS instruments. Aromatic hydrocarbons and alcohol mixtures were used to evaluate the system. Detection limits obtained ranged from 0.001 to 400 ng s^{-1} for a variety of elements including B, C, Br, Cl, Si, P, O, I, and S. In addition, elemental and isotopic ratios were determined.

GC-MIP-MS. In contrast to ICP-MS, more studies have focused on the use of MIP-MS as a detector for GC. Some early investigations used conventional GC-MS instruments with an MIP source. Markey and Abramson (28) described a microwave discharge interface with a conventional GC-MS instrument to determine ^{14}C containing compounds, often used as markers in metabolism experiments. A low-pressure MIP interface was used to convert all carbon compounds to CO and CO_2 which could be detected in a mass spectrometer.

Another study employing a conventional GC-MS instrument was described by Heppner (29). Both a H_2 and He low-pressure MIP were interfaced to the MS to determine C, O, N, S, and Cl in compounds separated by GC. Organic compounds were found to form simple, neutral molecules such as CO, HCN, and CH_4 which allowed elemental composition information of the original molecule to be determined. Elemental C, N, O, S, and Cl and C/O and C/N ratios in a variety of compounds were determined. The detection limit for carbon was found to be 0.3 μg s[-1].

The potential for determining elements in gas phase samples with a He MIP-MS instrument was demonstrated by Satzger and co-workers (*30*) and by Brown et al. (*31*). Detection limits for Br, Cl, and I were determined to be in the low pg s[-1] range in both studies. Background species present at low masses were also discussed in both publications.

Mohamad and co-workers (*32*) reported on the detection of halogenated hydrocarbons with an atmospheric pressure GC-MIP-MS system. A variety of halogenated hydrocarbons were separated and low pg detection limits were obtained. Atmospheric entrainment was limited via the use of a smaller sampler orifice.

Detection of halogenated compounds was also investigated by Creed et al. (33). An atmospheric-pressure He MIP-MS coupled with capillary GC resulted in pg sensitivity for Br and I; however, elevated backgrounds at m/z 35 and 37 complicated the detection of Cl-containing compounds. A low-pressure He MIP was interfaced to the mass spectrometer in order to eliminate atmospheric entrainment and reduce the low mass background. It was concluded that the use of the low-pressure system increased the sensitivity for the halogens. A study of helium purity showed that the use of research grade helium further reduced background species.

These studies with GC-MIP-MS have illustrated the potential of the technique to determine trace elements with species selectivity. Both atmospheric and low-pressure plasma systems have been investigated. In addition to halogenated compounds, the use of the technique can be extended to include other types of compounds.

Organotin Speciation using GC MIP-MS. Currently there is much interest in the determination of organotin compounds in the environment. These compounds are used commonly in catalysts and thermal stabilizers and can be highly toxic towards aquatic organisms and mammals (*34*). Reviewers (*34,35*) have addressed approaches to analyzing for organotin compounds and reached the conclusion that very sensitive analyses along with speciation information are requirements for accurately assessing the impact organotins have on the environment. The most common method of organotin speciation involves GC. With the sensitivity of plasma MS and the ease of interfacing GC with He MIP-MS, a study of this coupled technique for the determination of organotins is appropriate.

Suyani and co-workers have used He MIP-MS for the detection of several organotin compounds separated by GC (*36*). A VG PlasmaQuad ICP-MS was modified for use with the MIP by replacing the ICP torchbox assembly with the MIP cavity. A schematic of a typical atmospheric-pressure MIP system is shown in

Figure 1. The plasma was formed in a modified Beenakker TM_{010} cavity(31). An additional vacuum port was added to the first stage of the instrument to maintain adequate first stage pressure.

Two different torch designs were investigated. The first was a demountable tangential flow torch with a heated aluminum insert; an 8 mm quartz discharge tube was employed to reduce air entrainment. A separation of tetravinyltin (TVT), tetraethyltin (TET), and tetrabutyltin (TBT) using this torch configuration resulted in detection limits ranging from 1 - 5 pg Sn.

A second torch design was investigated in an attempt to improve the low signal intensity thought to be caused by diffusion of the tin analyte throughout the plasma. In order to minimize diffusion, a 3.5 cm long Ta injector tube was added to the standard torch design (37). The same separation of tetraorganotin compounds is shown in Figure 2 using the Ta injector (peaks A,B, and C correspond to TVT, TET, and TBT respectively; peaks D, E, and F are thought to be triorganotin impurities). The signal-to background ratio has improved by a factor of 4. Table I summarizes the figures of merit for the separation of six organotin compounds. Sub-pg detection limits are obtained, one order of magnitude lower than the torch without the Ta injector.

Table I. Analytical Figures of Merit Using the Ta Injector Torch[a]

Compound[b]	TVT	TET	TET-Br	TPrT-Cl	TBT	TBT-Cl
RSD[c], %	4.3	4.0	4.2	3.7	2.4	4.0
Log plot slope	0.991	0.958	0.972	0.980	0.983	0.953
LDR[d] (decades)	3	3	3	3	3	3
LOD[e] (pg of Sn)	0.35	0.12	0.18	0.13	0.09	0.17

[a] Adapted from ref. 35.
[b] TET-Br (triethyltin bromide), TPrT-Cl (tripropyltin chloride), TBT-Cl (tributyltin chloride)
[c] Ten replicate injections of 50 pg of Sn for each compound.
[d] Linear dynamic range
[e] Detection limit defined as 3(sigma) of background cps/slope of calibration curve

Detection of Non-metal GC Effluents with Low-Pressure MIP-MS. Many low mass elements such as phosphorus, sulfur, and chlorine are of interest because of their presence in pesticides and herbicides, however, atmospheric-pressure GC-MIP-MS systems have some limitations with regard to analyzing for many of these low-mass elements. Ultratrace detection of the halogens, phosphorus and sulfur can be hindered by isobaric interferences from polyatomic ions

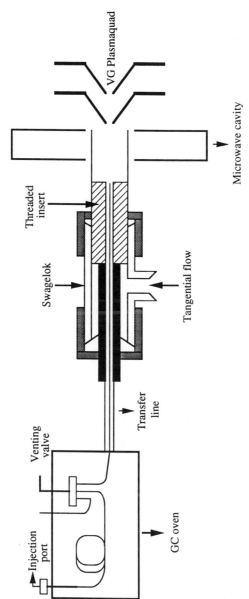

Figure 1. Interface of gas chromatograph to a helium MIP-MS. Reproduced with permission from ref. 33. Copyright 1988 The Royal Society of Chemistry.

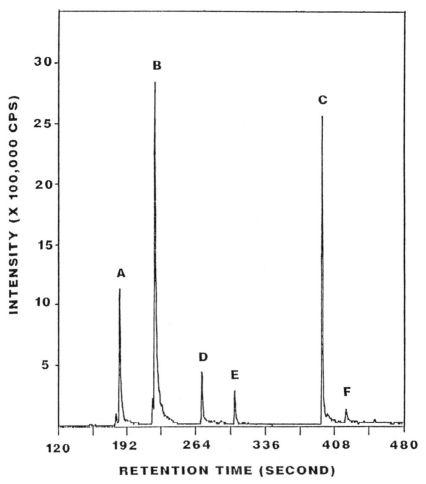

Figure 2. Chromatogram of a mixture of tetraorganotin compounds obtained at m/z = 120 with Ta injector torch. A, TVT; B. TET; and C, TBT; D, E, F, see text. Adapted from ref. 36.

formed from gas impurities and atmospheric entrainment (*12,32*) as well as system impurities of the elements themselves.

One approach that has been taken to minimize these low-mass polyatomics is the use of a low-pressure plasma interface (*33*). Along with eliminating the atmosphere as a source of polyatomic interferences, because of low flow rates used, a low-pressure plasma can make the use of research grade helium more affordable thereby decreasing the levels of He gas impurities (*33*).

Halogenated Hydrocarbons. Creed et al. (*38*) have used a low-pressure He MIP-MS as a detector for gas chromatography. A schematic diagram of the low-pressure interface is shown in Figure 3. A vacuum fitting was silver-soldered to a nickel sampling cone making a low-pressure seal to the plasma discharge tube. A modified Beenakker TM_{010} cavity was used. The cavity and torch assembly replaced the standard ICP torchbox on a VG PlasmaQuad instrument.

Effects of power and plasma gas flow rate (indicated by the first stage pressure) were studied. Signal intensity at mass 127 (iodine) and mass 129 (xenon as a helium impurity) was monitored. Changes in the plasma plume size explained decreasing signal with increasing pressure and increasing signal with increasing power.

Figure 4 illustrates a typical chromatogram obtained with the low-pressure He MIP for a 1 pg injection of iodobenzene. The background shift at 120 s was a result of solvent venting. The detection limit for iodine was calculated to be 0.1 pg. Chromatograms were also obtained for bromononane and chlorotoluene. The detection limit for Br, calculated to be 3.5 pg, was elevated due to background interference from $^{61}NiO^+$ and $^{60}NiOH^+$ resulting from ablation of the Ni sampling cone. Detection of Cl was limited due to extremely high backgrounds at masses 35 and 37 with a detection limit of 24 pg.

In an attempt to lower the background at masses 35 and 37 the sampling interface was machined as a one piece aluminum sampler in order to eliminate the silver solder (a potential Cl source). The background species present were then investigated. Isotope ratios of Cl background (35/37) were in error by only 2.3% indicating that some residual Cl is present in the system. Further reductions in background species could potentially extend the detection limits for Br and Cl.

Phosphorus and Sulfur. Although AES has excellent sensitivity for P and S, the improved sensitivity of MS for metals prompts an investigation with GC MIP-MS. Detection of P and S is difficult with an atmospheric-pressure plasma due to isobaric interferences from NOH^+ (m/z = 31) and O_2^+ (m/z = 32), respectively. These interferences, resulting from air entrainment, can be minimized with a low-pressure plasma.

Story and co-workers (*J. Anal. At. Spectrom.*, **1990**, *5*, in press) have used a low-pressure MIP-MS system to detect P and S compounds separated by GC. The

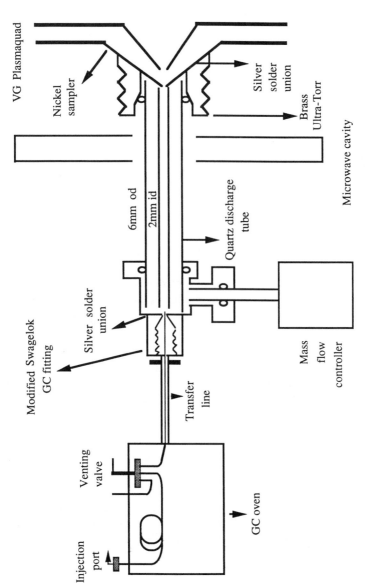

Figure 3. Low-pressure MIP torch interface to the PlasmaQuad. Reproduced with permission from ref. 38. Copyright 1990 The Royal Society of Chemistry.

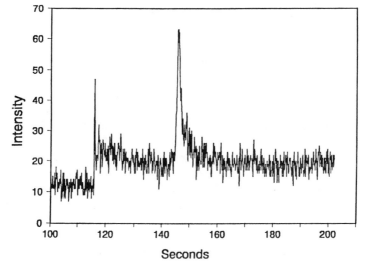

Figure 4. Chromatographic trace for a 1 pg injection of iodobenzene. Adapted from ref. 38.

low-pressure interface is similar to that described in Figure 3, utilizing the one-piece aluminum sampler.

Initially He was used as the plasma gas and two different torches were investigated. The first torch, a 1 mm i.d. quartz tube, performed poorly. The background observed at mass 31 was found to vary from 1000 to 10000 count s^{-1}. A 180 ng injection of triethyl phosphite produced a very small peak over this high background and a detection limit of 90 ng P. The low sensitivity was thought to be due to inefficient cooling of the torch resulting in a very hot quartz tube and subsequent reaction with P.

In an effort to improve the performance, a 4 mm i.d. quartz tube with an air cooling jacket was used. The signal intensity of the triethyl phosphite peak improved by a factor of 5. The resulting detection limit was improved to 1 ng P; however, inefficient cooling of the torch was still thought to limit sensitivity.

High backgrounds at mass 32 precluded the determination of sulfur compounds with a He plasma. A low-pressure nitrogen plasma was investigated for the ability to determine sulfur. The low ionization potential of N_2 (14.5 eV vs. 24.5 eV for He) results in a less energetic plasma that should decrease the formation of O_2^+ (12.06 eV for O_2) and allow the determination of S.

Two commercial pesticides, Malathion and Diazinon, were used to evaluate the performance of the N_2 MIP as a GC detector. Single ion chromatograms were obtained for the separation of these compounds by monitoring mass 32 for S and mass 31 for P. Two peaks with good resolution resulted for both chromatograms, however, the chromatogram obtained for P exhibited some peak broadening, providing further evidence for the reaction of P with the hot quartz torch. The resulting figures of merit for the N_2 MIP system are shown in Table II. The detection limits for P are 0.6 and 0.8 ng, while those for sulfur are 0.2 and 0.5 ng.

Table II. Figures of Merit for Nitrogen Plasma.[a]

Compound	Element	Slope Log-Log (1 ng - 500 ng)	Detection Limit (ng element)
Diazinon	P	1.08	0.79
Diazinon	S	1.03	0.51
Malathion	P	0.954	0.57
Malathion	S	0.893	0.15

[a]Adapted from Story et al., *J. Anal. At. Spectrom.* **1990**, *5*, in press.

Recently, the low-pressure He MIP has been used to obtain multielement chromatograms (*39*). A mixture of Cl-containing compounds, Diazinon and

Malathion, was used. Figure 5a shows the total ion beam chromatogram (sum of all masses from m/z = 31 to 37 including background ions) obtained for the mixture. Figure 5b shows the separate mass chromatograms obtained for ^{35}Cl, ^{37}Cl, and ^{31}P with peaks corresponding to the total chromatogram. The ability to obtain multielement information on eluting peaks can greatly extend the usefulness of plasma MS as a detector for chromatography.

The low-pressure MIP-MS system shows promise for the detection of P and S compounds separated by GC. Although better signal and detection limits are expected with the more energetic He plasma, high backgrounds and reactions with the quartz torch hinder its use. Thus far the use of a N_2 plasma has resulted in improved detection for P and S. With more efficient torch cooling plus the use of a reagent gas such as H_2, it may be possible to improve the He plasma results for P and S.

Plasma MS detection for HPLC

HPLC-ICP-MS. Element specific detection for HPLC is becoming more important as trace speciation studies increase. Atomic spectrometric techniques have been used as detectors for HPLC and the advantages of coupling the two have been the subject of several reviews (40-45). Emission (46-50) methods have been used as element specific detectors for HPLC and are the subject of other chapters in this book. Advantages of coupling ICP-AES with HPLC include multielement detection and the ability to obtain real-time chromatograms. Unfortunately, ICP-AES does not always provide the sensitivity necessary for trace element speciation of real samples. The greater sensitivity of ICP-MS over ICP-AES can provide this ability.

A number of publications have described the coupling of HPLC and ICP-MS and several have been reviewed in more detail (Heitkemper, D.T.; Caruso, J.A. *Trace Metal Analysis and Speciation*, in press). Metal speciation has been the subject of several publications. Hg speciation was studied by Bushee (51); methyl mercury was determined in an NIST reference material and thimerosal in contact lens solution. Crews et al. (52) studied the speciation of cadmium in pig kidney with ICP-MS using size exclusion chromatography. Using ICP-MS, Cd species at normal "background" levels were studied. The majority of soluble Cd in the kidney tissue was found to be associated with a metallothionein-like protein that survives both cooking and digestion.

Suyani (53) and co-workers used HPLC-ICP-MS to detect organotin species. Ion exchange and ion pair chromatography were used to obtain detection limits ranging from 0.4 to 1 ng for several triorganotin species. In a separate publication, Suyani et al. (54) separated a series of 5 organotin species using micellar liquid chromatography coupled with ICP-MS. Using sodium dodecyl sulfate as mobile phase and a C18 stationary phase, pg detection limits were obtained.

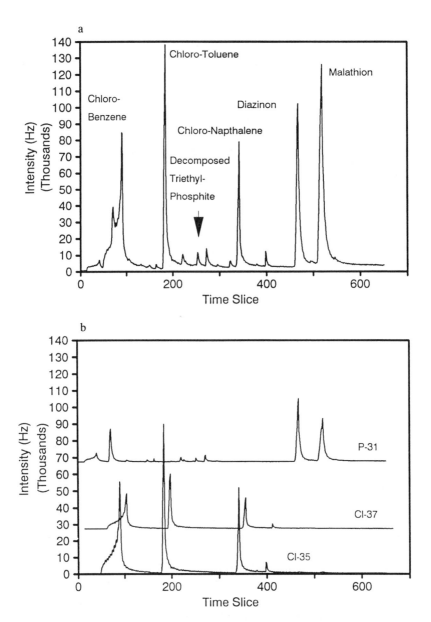

Figure 5. Multielement chromatograms of Cl and P, a)total ion beam, b) separate single ion chromatograms. Adapted from ref 39.

Matz, Elder and co-workers (55) have used HPLC ICP-MS to determine gold metabolites in gold based anti-arthritis drugs. A detailed description is found in a separate chapter of this volume.

Many of the metal speciation studies focus on arsenic. The widespread use of As compounds in a variety of industrial and agricultural applications results in a need for methods that can accurately assess exposure to toxic As compounds. In addition to the initial publication by Thompson and Houk (56), three other publications (57-59) also investigate As speciation. Beauchemin et al. (57) quantified As species in a dogfish muscle reference material and concluded that the major As species in the material was arsenobetaine, totalling 84%. In an additional publication, Beauchemin and co-workers (58) separated As(III), As(V), monomethylarsonic acid (MMA), dimethylarsinic acid (DMA) and arsenobetaine (AsB) with various forms of HPLC detected by ICP-MS. Both ion pairing and ion exchange chromatography were used with detection limits ranging from 50-300 pg.

Heitkemper and co-workers (59) have also used HPLC-ICP-MS for As speciation. They used anion exchange HPLC and calculated detection limits for the 4 species (As(III), As(V), MMA and DMA) ranging from 20-91 pg for aqueous samples and 36-96 pg for urine samples. The determination of As(III) in urine was hindered by the presence of an interference at mass 75. This interference was explained by the co-elution of Cl-containing species which can form $^{40}Ar^{35}Cl^+$ species.

Ion Chromatography ICP-MS for Interference Elimination. Ion chromatography(IC) is attractive for element speciation because it can separate inorganic and organic charged species as well as free ions. Coupling IC and ICP-MS can provide a powerful tool for trace element speciation because IC does not use organic solvents that can potentially destabilize the plasma and elevate backgrounds. In addition, single column IC uses low buffer concentrations which reduces matrix effects and salt buildup at the plasma interface.

Sheppard et al. (*J. Anal. At. Spectrom.* **1990**, *5*, in press.) have used IC to resolve chloride from As species chromatographically and thus eliminate the interference in ICP-MS from $ArCl^+$ at m/z = 75 which was reported by Heitkemper et al. (59) when speciating As in urine samples. Urine is an important tool in measuring exposure to As because it is the major elimination pathway in humans. Urine is approximately 0.15 M in NaCl which is 10^5 times more concentrated than any arsenic species assuming normal As concentrations. The elimination of this interference in ICP-MS would aid in the trace detection of As species in urine.

Initially the presence of a chloride interference at m/z = 75 was confirmed by injection of 1% NaCl. Two peaks appeared in the chromatograms obtained at masses 51 ($^{35}Cl^{16}O^+$), 75 ($^{40}Ar^{35}Cl^+$), and 77 ($^{40}Ar^{37}Cl^+$). The isotope ratio for chlorine (m/z = 75/77) obtained for both peaks was approximately correct (3:1). The lack of any peaks at mass 82 ($^{82}Se^+$) was further evidence that the peak at mass 77 was due to $ArCl^+$ and not Se.

Arsenite (AsIII)) and arsenate (As(V)) were separated, the two methylated species, MMA and DMA being partially resolved from the As(III). The effect of the NaCl concentration on the separation of As(III) and As(V) was examined. At low NaCl concentrations (0.01%) the interference appeared as a shoulder on the As(V) peak. By increasing the concentration to 1% NaCl the As(V) peak was no longer discernable. This effect was explained by the masking of the ion exchange sites in the column which cause the As(V) peak to split and partially co-elute with the As(III) peak. Because the concentration of Cl in urine is on the order of 1%, the interference of $ArCl^+$ on As in urine can only be resolved if an appropriate dilution factor is used. A 20-fold dilution of the urine was used representing less than 0.05% chloride. An example chromatogram showing the separation of $ArCl^+$ from As(III) and As(V) in urine is illustrated in Figure 6. Table III shows the resulting figures of merit for the determination of As in urine. The detection limits calculated for the As species ranged from 0.3 - 0.7 ng in 100% urine.

Table III. Figures of Merit in 1:20 Urine:Water.[a]

	Retention Time (min)	LOD (ppb)	Absolute LOD (ng)	LDR	%RSD[b]
As(III)	2.7	0.17	0.017	> 2 orders	6.0%
As(V)	5.3	0.35	0.035	> 2 orders	4.5%
DMA[c]	3.2	0.21	0.021	ND[d]	5.5%

[a]Adapted from Sheppard et al. *J. Anal. At. Spectrom.* **1990**, *5*, in press.
[b]Ten replicate injections of 10 ppb standard addition
[c]Based on 100 ppb DMA standard addition
[d]ND = Not Determined

The accuracy of the method was tested by analyzing three freeze-dried urine standards by the method of standard additions. The total concentration of all As species was compared with the reported values and found to be in good agreement thus illustrating that the interference from $ArCl^+$ is sufficiently minimized.

HPLC-MIP-MS. The advantages of using a helium MIP, including improved detection for the halogens, were discussed previously. The primary disadvantage, however, is the inability of the plasma to tolerate aerosol introduction. The resulting solvent load can destabilize and extinguish the plasma. Few reports of aqueous sample introduction into the He MIP are available (*60-64*). Recently Creed and co-workers (*65*) have described a He MIP-MS capable of handling aqueous sample introduction. They used a moderate power He MIP with a He MAK nebulizer (*66*).

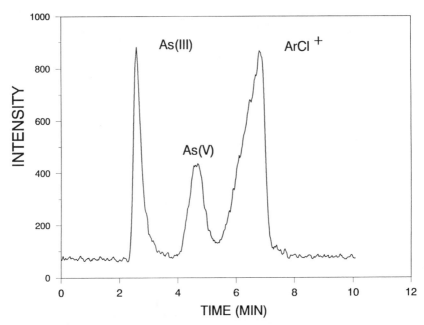

Figure 6. Chromatogram of 100 ppb As(III) and As(V) in 1:20 urine showing separation from ArCl$^+$ interference at m/z = 75. Reproduced with permission from Sheppard, et al., *J. Anal. Atomic Spectrom.*, **1990**, *5*, in press. Copyright 1990 The Royal Society of Chemistry.

The addition of N_2 to the auxiliary He flow was necessary to improve the linear dynamic range for metals. Detection limits for Br, I and several metals were in the sub-ppb range while those for Cl were in the low-ppb range due to relatively high backgrounds at m/z = 35 and 37.

Relatively few attempts (67-69) have been made to couple HPLC to He MIP-AES despite the need for a nonmetal element-specific HPLC detector. Heitkemper et al. (*70*) have investigated the potential of coupling HPLC to He MIP-MS for the determination of halogenated organic compounds.

Initial experiments involved direct organic solvent introduction with a He MAK nebulizer. Methanol, ethanol, acetonitrile and 2-propanol were introduced into the system in concentrations of 5 to 100%. The introduction of acetonitrile at concentrations over 5% caused excessive carbon buildup on the sampling cone. Solutions containing 70% methanol, ethanol, or 2-propanol caused no visible carbon buildup over a 30 min time period. The background mass spectrum was also determined as a function of the alcohol concentration. Because backgrounds increase significantly with increasing alcohol concentration the amount of organic modifier must be minimized. Detection limits for As, Cd, Sn, and Pb in 20% methanol were calculated to be 1 ppb for all four elements.

Experiments using HPLC-MIP-MS were performed with a concentric nebulizer designed for use with He. A typical chromatogram obtained with this system is shown in Figure 7 where dibromomethane has been separated from

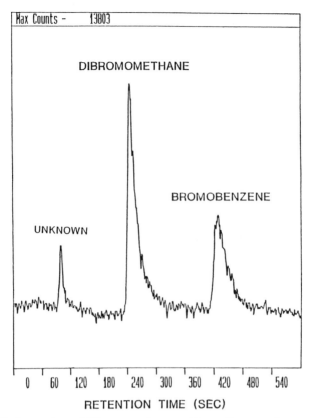

Figure 7. Separation of dibromomethane and bromobenzene by reversed-phase HPLC with He MIP-MS detection. Each peak corresponds to 1 ng Br. Reproduced with permission from ref. 70, J. Chromatog. Sci. Copyright 1990 Preston Publications, A Division of Preston Industries, Inc.

bromobenzene. Chromatograms for the separation of ethyliodide and iodobenzene as well as for chlorobenzene were also obtained. Figures of merit for these compounds are shown in Table IV, pg levels of detection being obtained for Br and I compounds. The determination of chlorine was hindered by large backgrounds.

Additionally, the He MIP-MS system was evaluated as a method of eliminating the ArCl$^+$ interference on the determination of As. As was determined by solution nebulization rather than with the HPLC in place. The detection limit calculated, 0.1 ppb, is comparable to that obtained with ICP-MS. The effect of the interference was determined by introducing a 5% HCl solution. The signal equivalent to a concentration of 0.34 ppb As was found at m/z 75, resulting from Ar contamination in the plasma (Ar is a known contaminant of He gas). In a comparable study, using ICP-MS, the resulting signal is equivalent to 6 ppb As (71).

Table IV.   Analytical Figures of Merit for Halogenated
Hydrocarbons by HPLC-He-MIP-MS[a]

Compound[b]	A	B	C	D	E
m/z monitored	79	79	127	127	35
retention time(min)	7.0	4.1	9.5	5.3	6.5
LOD (pg as element)[c]	63	25	0.9	1.4	6300
LDR (decades)	3	3	4	4	1
log plot slope	0.98	0.97	0.96	0.98	1.01
reproducibility (% RSD, n ≥ 6)	6.6	5.2	8.5	8.3	6.5

[a]Reproduced with permission from ref. 70, J. Chromatog. Sci. Copyright 1990
  Preston Publications, A Division of Preston Industries, Inc.
[b]Compound A = bromobenzene, B = dibromomethane, C = 1-iodobutane,
  D = ethyliodide, E = chlorobenzene
[c]Detection limit defined as 3(sigma) of background cps/slope of calibration curve

Conclusions and Future Development

As evidenced by the increasing number of publications, interest in chromatographic
detection by plasma MS is growing. However, further work is necessary.
Multielement detection and isotope ratios on eluting peaks is one area of interest.
Additionally, speciation studies involving real samples for many elements should be
investigated. Finally, the use of alternative plasma sources, i.e. helium, nitrogen,
and mixed gas plasmas and alternative separation techniques, i.e. supercritical fluid
chromatography, will be an area of future research.

Acknowledgments

The authors are grateful to the National Institute of Environmental Health Sciences
for support under grant number ES03221 and to the University of Cincinnati,
University Research Council for funding.

Literature Cited

1.   Houk, R.S. *Anal. Chem.* **1986**, *58*, 97A-105A.
2.   Houk, R.S.; Thompson, J. J. *Mass Spectrom. Rev.* **1988**, *7*, 425-461.
3.   Gray, A.L. In *Inorganic Mass Spectrometry*; Adams, F.; Gijbels, R.;
     VanGrieken, R., Ed.; Wiley: New York, 1988, Ch. 6.
4.   Horlick, G.; Tan, S.H.; Vaughan, M.A.; Shao, Y. In *Inductively Coupled
     Plasmas in Analytical Atomic Spectrometry*; Montaser, A.; Golightly, D.W. ,
     Eds.; VCH Publishers: New York, 1987, pp 361-398.

5. Gray, A.L. *Spectromchim. Acta*, **1985**, *40B*, 1525.
6. Douglas, D.J.; Houk, R.S. *Prog. Anal. At. Spectrosc.* **1985**, *8*, 1.
7. Date, A.R.; Gray, A.L. *Application of Inductively Coupled Plasma Mass Spectrometry*, Blackie: Glasgow & London, 1989.
8. Douglas, D.J.; French, J.B. *Anal. Chem.* **1981**, *53*, 37.
9. Douglas, D.J.; Quan, E.S.K.; Smith, R.G. *Spectrochim. Acta* **1983**, *38B*, 39.
10. Koppenal, D.W.; Quinon, L.F. *J. Anal. At. Spectrom.* **1988**, *3*, 667.
11. Montaser, A.; Chan, S.; Koppenaal, D.W. *Anal. Chem.* **1987**, *59*, 1240.
12. Tan, S.H.; Horlick, G. *Appl. Spectrosc.* **1986**, *40*, 445.
13. Mohamad, A.H.; Caruso, J.A. In *Advances in Chromatography*; Giddings, J.C., Ed.; Marcel Dekker, Inc.: New York, 1987, Vol. 3; pp 191-227.
14. Fior, R.L.; *Am. Lab.* **1989**, *21*, 40.
15. Ebdon, L.; Hill, S.; Ward, R.W. *Analyst* **1986**, *111*, 1113.
16. Windsor, D.L.; Denton, M.B. *Appl. Spectrosc.* **1978**, *32,* 366.
17. Windsor, D.L.; Denton, M.B. *Anal. Chem.* **1979**, *51*, 1116.
18. Windsor, D.L.; Denton, M.B. *J. Chromatogr. Sci.* **1979**, *17*, 492.
19. Brown, R.M. Jr.; Fry, R.C. *Anal. Chem.* **1981**, *53*, 532.
20. Brown, R.M. Jr.; Northaway, S.J.; Fry R.C. *Anal. Chem.* **1981**, *53*, 934.
21. Keane, J.M.; Brown, D.C.; Fry, R.C. *Anal. Chem.* **1985**, *57*, 2526.
22. Bruce, M.L.; Caruso, J.A. *Appl. Spectrosc.* **1985**, *39*, 942.
23. Haas, D.L.; Caruso, J.A. *Anal. Chem.* **1985**, *57*, 846.
24. Luffer, D.R.; Galente, L.J.; David, P.A.; Novotny, M.; Hieftje, G.M. *Anal. Chem.* **1988**, *60*, 1365.
25. Goode, S.R.; Chambers, B.; Buddin, N.P. *Appl. Spectrosc.* **1983**, *37*, 439.
26. Evans, J.C.; Olsen, K.B.; Sklarew, D.S. *Anal. Chim. Acta* **1987**, *194*, 247.
27. Chong, N.S.; Houk, R.S. *Appl. Spectrosc.* **1987**,*41*, 66.
28. Markey, S.P.; Abramson, F.P. *Anal. Chem.* **1982**, *54*, 2375.
29. Heppner, R.A. *Anal. Chem.* **1983**, *55*, 2170.
30. Satzger, R.D.; Fricke, F.L.; Brown, P.G.; Caruso, J.A. *Spectromchim. Acta* **1987**, *42B*, 705.
31. Brown, P.G.; Davidson, T.M.; Caruso, J.A. *J. Anal. At. Spectrom.* **1988**, *3*,763.
32. Mohamad, A.H.; Creed, J.T.; Davidson, T.M.; Caruso, J.A. *Appl. Spectrosc.* **1989**, *43*, 1127.
33. Creed, J.T.: Mohamad, A.H.: Davidson, T.M.; Ataman, G.; Caruso, J.A., *J. Anal. At. Spectrom.* **1988**, *3*, 923.
34. Blair, W.; Olsen, G.; Brinckman, F.; Paule, R.; Becker, D. *An International Butyltin Measurements Methods Intercomparison:*... NBSIR 86-3321, Nat'l Bureau of Standards, Dept. of Commerce, Feb. 1986.
35. Thompson, J.; Sheffer, M.; Pierce, R.; Chau, Y.; Cooney, J.; Cullen, W.; Mcquire, R. *Organotin Compounds in the Aquatic Environment:*... NRCC #22494, National Research Council, Canada, Assoc. Comm. Environ. Safety, 1985.

36. Suyani, H.; Creed, J.; Caruso, J.A.; Satzger, R.D. *J. Anal. At. Spectrom.* **1989**, *4*, 777.
37. Satzger, R.D.; Brueggemeyer, T.W. *Mikrochim. Acta* **1989**, *III*, 239.
38. Creed J.T.; Davidson, T.M., Shen, W.L.; Caruso, J.A. *J. Anal. At. Spectrom.* **1990**, *5*, 109
39. Sheppard, B.S.; Shen, W.; Story, W.C.; Caruso, J.A. Biodegradation of Hazardous Waste Conference Logan, Utah, April, 1990.
40. Van Loon, J.C. *Anal. Chem.* **1979**, *51*, 1139A.
41. Krull, I.S. *Trends in Analytical Chemistry*, **1984**, *3*, 76.
42. Irgolic, K.J.; Brinckman, F.E. Dahlou Conference Report, Berlin, Sept. 2-7, 1984, Springer Verlag, 1986.
43. Uden, P.C. *Trends in Analytical Chemistry* **1987**, *6*, 238.
44. Gardiner, P.E. *J. Anal. At. Spectrom.* **1988**, *3*, 163.
45. Cappon, C.J. *LC/GC* **1988**, *6*, 584.
46. McCarthy, J.P.; Caruso, J.A.; Fricke, F.L. *J. Chromatogr. Sci.* **1983**, *21*, 389.
47. Irgolic, K.J.; Stockton, R.A.; Chakraborti, D.; Beyer, W. *Spectrochim. Acta* **1983**, *38B*, 437.
48. Bushee, D.S.; Krull, I.S.; Demdo, P.R.; Smith, S.B. Jr. *J. Liq. Chromatogr.* **1984**, *7*, 861.
49. Gast, C.H.; Kraak, J.C.; Poppe, H.; Maessen, F.J.M.J. *J. Chromatogr.* **1979**, *185*, 549.
50. Krull, I.S.; Panaro, K.W. *Appl. Spec.* **1985**, *39*, 960.
51. Bushee, D.S. *Analyst* **1988**, *113*, 1167.
52. Crews, H.M.; Dean, J.R.; Ebdon, L.; Massey, R.C. *Analyst* **1989**, *114*, 895.
53. Suyani, H.; Creed, J.; Davidson, T.; Caruso, J. *J. Chromatogr. Sci.* **1989**, *27*, 139.
54. Suyani, H.; Heitkemper, D.; Creed, J.; Caruso, J. *Appl. Spectros.* **1989**, *43*, 777.
55. Matz, S.G.; Elder, R.C.; Tepperman, K. *J. Anal. Atom. Spectrom.* **1989**, *4*, 767.
56. Thompson, J.J.; Houk, R.S. *Anal. Chem.* **1986**, *58*, 2541.
57. Beauchemin, D.; Bednas, M.E.; Berman, S.S.; McLaren, J.W.; Siu, K.W.M.; Sturgeon, R.E. *Anal. Chem.* **1988**, *60*, 2209.
58. Beauchemin, D.; Siu, K.W.M.; McLaren, J.W.; Berman, S.S. *J. Anal. Atom. Spectrom.* **1989**, *4*, 285.
59. Heitkemper, D.; Creed, J.; Caruso, J.; Fricke, F. *J. Anal. At. Spectrom.* **1989**, *4*, 279.
60. Haas, D.L.; Caruso, J.A. *Anal. Chem.* **1984**, *56*, 2014.
61. Michlewicz, K.W.; Carnahan, J.W. *Anal. Chem.* **1985**, *57*, 1092.
62. Michlewicz, K.W.; Carnahan, J.W. *Anal. Chem.* **1986**, *58*, 3122.
63. Brown, P.G. Ph.D. Dissertation, University of Cincinnati, 1988.
64. Stahl, R.G.; Timmins, K.J. *J. Anal. At. Spectrom.* **1987**, *2*, 557.

65. Creed, J.T.; Davidson, T.M.; Shen, W.L.; Brown, P.G.; Caruso, J. A. *Spectrochim. Acta* **1989**, *44B*, 409.
66. Meddings, B.; Kaiser, H.; Anderson, H. *Proc. Intl. Winter Conf. Spectrochem. Anal.* San Juan, Puerto Rico, 1980.
67. Billet, H.A.H.; van Dalen, J.P.J.; Schoenmakers, P.J.; De Galan, L. *Anal. Chem.* **1983**, *55*, 847.
68. Michlewicz, K.G.; Carnahan, J.W. *Anal. Lett.* **1987**, *20*, 1193.
69. Zhang, L.; Carnahan, J.W.; Winans, R.E.; Neill, P.H. *Anal. Chem.* **1989**, *61*, 895.
70. Heitkemper, D.; Creed, J.; Caruso, J.A. *J. Chromatogr. Sci.* **1990**, *28*, 175.
71. Hutton, R.C. In *VG Isotopes Technical Information*, VG Isotopes, Winsford, UK, 1986.

RECEIVED April 12, 1991

Chapter 18

Element-Specific Detection of Metallodrugs and Their Metabolites

High-Performance Liquid Chromatography—Inductively Coupled Plasma Mass Spectrometry

R. C. Elder, W. B. Jones, and Katherine Tepperman

Biomedical Chemistry Research Center, University of Cincinnati, Cincinnati, OH 45221–0172

Many of the growing number of metal-based drugs in use or testing, are metabolically active. To separate and characterize both the drugs and their metabolites we use the element-specific detection capabilities of an inductively coupled plasma mass spectrometer interfaced to a high performance liquid chromatograph. Detection limits are good, typically in the 1-10 ppb range for chromatography and ca. three orders of magnitude better than for atomic absorption spectrometry. Examples of size exclusion, reversed phase and weak anion exchange chromatography are given showing multielement detection from a single sample injection. These samples are from patients undergoing gold drug therapy for rheumatoid arthritis. Problems associated with the use of organic solvents are illustrated and solutions to these problems are demonstrated.

Recently metal-based drugs have attracted a great deal of interest. Cis-Platin, diamminedichloroplatinum(II), is widely used for the treatment of testicular, ovarian and bladder cancers and is considered to be curative (1). Gold sodium thiomalate is used to treat several hundred thousand sufferers from rheumatoid arthritis with weekly injections of "gold" (2). Very recently gold thioglucose has been shown to inhibit the reverse transcriptase of the human immunodeficiency virus (HIV) and proposed as a possible treatment for AIDS patients (3). Silver sulfadiazine is used as an antibacterial for burn patients (4) and tartar emetic, potassium antimony tartrate, has been used since ca. 1600 in the treatment of parasitic disorders such as leismaniasis and schistosomiasis (5) but is more frequently used in veterinary medicine than for treatment of humans.
Other pharmaceuticals utilize metal radionuclides either as diagnostic imaging agents or as therapeutics. Thus ^{99m}Tc has been used to synthesize bone (6) and heart

0097–6156/92/0479–0309$06.00/0

(7) imaging agents. These materials depend on the propensity of a particular metal complex to concentrate in a tissue of interest. There they emit gamma rays which can generate an image of the region of localization. Thus Tc(HEDP), where HEDP stands for hydroxyethylidene diphosphonate, is a bone imaging agent which specifically binds to growing bone. It may be used to image the healing process for broken bones or to image bone tumors. If this agent is modified by the replacement of ^{99m}Tc by ^{186}Re the complexes behave similarly, concentrating for example on the bone tumors associated with metastatic breast or prostate cancer (8). However, ^{186}Re is both a beta and a gamma emitter. The gamma rays may be used to image the tumors; but, more importantly, the beta rays travel only a short distance before giving up their energy to the surroundings thus causing local tissue damage or cell death. When localized in a tumor these compounds thus become therapeutic. While it is true that element-specific separation and characterization of these materials can be monitored via their radioactive decay, it is also possible to perform experiments, in the case of rhenium complexes, with the non-radioactive analogues. This avoids the difficulties associated with handling radioactive materials while performing exploratory studies. These are but a few examples of the metal-based pharmaceuticals in use or in development. Since these drugs all seem to rely on the metals for their special properties, it is clearly important to follow them and study their interactions within the biological milieu in an element-specific manner. We reviewed the area of metallodrugs in 1987 (9) from a somewhat different perspective.

One of the paradigms of modern drug design is that of the structure activity relationship, SAR, that is, by making fine modifications in the structure of a drug we are able to tune its activity or decrease its side effects. Of course, to utilize the SAR we must know the composition(s) and structure(s) of the active form(s) of the drugs. This is made more complicated by the fact that many of the metal drugs mentioned above are readily metabolized and it is a metabolite which is the active form of the drug. Thus we need to separate and characterize these materials in order to improve efficacy or reduce toxicity. For the last several years our group has been exploring the use of an inductively coupled plasma mass spectrometer (ICP-MS) as an element-specific detector for high performance liquid chromatography (HPLC). Others have also utilized ICP-MS as an element-specific detector, notably our colleague J. Caruso and his students. They give a broad review of the field in the previous chapter of this volume. We rather concentrate on the problems which are encountered working in the biomedical, area dealing with the separation and characterization of metallodrugs and their metabolites.

Most of our work has been with the gold-based drugs used to treat rheumatoid arthritis (RA). There are two such drugs in widespread use. Gold sodium thiomalate (myochrysine) is given by injection on a weekly basis and has been in use for over sixty years. Triethylphosphine-gold(I)tetraacetylthioglucose (auranofin) is an orally administered drug which has been available for only the past few years (2). RA is a disease whose cause is unknown. It is characterized by painful swelling of the joints, thinning of the synovial fluid (which normally lubricates the joints) and eventual erosion of the cartilage which provides the bearing surface of the joints. Standard treatment which involves regular, high-level doses of aspirin reduces inflammation and relieves pain. It does not deter cartilage erosion, however. RA is a progressively debilitating disease and patients can be afflicted for thirty or forty years. There is no known cure, but the gold drugs have been shown in controlled, double blind studies to cause remission.

Early Element Specific Detection Experiments

Two experiments we performed led us to seek element-specific detection for HPLC. First, we performed an experiment (10) using a sac of everted hamster gut showing that the drug auranofin appeared to be metabolized before uptake in the intestine. Thus, the absorbed form of the drug was unlikely to be that administered but rather some chemically changed material. Second, Susan Matz, at the time a postdoctoral fellow in our group, showed as indicated in Figure 1, that different patients metabolize the same drug differently. Since gold was clearly the critical component of these metabolites (the ligands without the metal are not effective in treating RA), we needed to develop gold-specific detection to follow the metabolites of interest. Our initial efforts utilized an atomic absorption spectrometer (AAS) with a gold lamp as an element-specific detector. The difficulty was with the detection limit. AAS provides a detection limit of ca. 5.0 ppm, after peak spreading and dilution occurring during chromatographic separation are taken into account. Since total gold levels in the blood or urine of patients on gold drugs are in the range of 1.0 to 0.1 ppm, obviously, a lower detection limit was desirable. Table I provides a comparison of various chromatographic detectors. UV detectors have no element specificity; AAS and ICP/AES provide element-specific detection but do not have the necessary detection limits to be used for samples with no preconcentration step.

Table I. Chromatograhic Detectors

Detector	Gold Drug HPLC Sensitivity	Element Specificity	Multi Element	Price
uv 254 nm	10.0 ppm	none	n.a.	$ 2,000
uv 214 nm	0.1 ppm	none	n.a.	$ 4,000
AAS	5.0 ppm	yes	single	$ 20,000
ICP/AES	0.5 ppm	yes	multi	$ 60,000
ICP/MS	5.0 ppb	yes	multi	$200,000

(Note that the samples in Figure 1 were concentrated on a SepPak precolumn by a factor of twenty). ICP-MS clearly has the needed detection limit although its cost is extremely high. Other techniques which have been

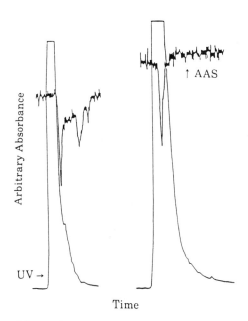

Figure 1. C18 Chromatography of patient urine. Samples from two auranofin patients were concentrated using a SepPak and chromatographed on a C18 column. The lower trace represents the UV 214 signal, while the upper trace, on a slight delay, represents the gold signal from the AA.

suggested, such as graphite furnace vaporization of samples coupled with AAS or atomic emission spectrometry (AES), are not well suited to use with a flowing stream of eluent from a chromatographic column.

The ICP-MS which we use as a detector is a SCIEX Elan 250. We have described the experimental conditions elsewhere (11). Basically the eluent from the HPLC is plumbed directly to the nebulizer of a low flow Meinhard nebulizer on the standard spray chamber and plasma torch of the ICP-MS. Standard flow rates are 1.0 mL/min and are easily compatible with both systems.

Flow Injection Analysis

Frequently with medical samples the volumes are restricted to small amounts. We have used flow injection analysis with an injection valve from the chromatographic system. This is well suited to blood analyses where we have much less material than is optimal for normal flow analyses. We have encountered problems, however, due to the small amounts of analyte species present.

Figure 2 shows results of FIA on Cu, Zn and Cd standards injected under two different sets of conditions. Note, that single injections were made of a mixture of all three standard materials. The Sciex mass spectrometer may be set to "peak hop" to the various mass/charge ratios for which analysis is desired. Dwell times at a particular m/e ratio as short as 30 msec are practical. Thus these and subsequently depicted chromatograms are in effect simultaneous multielement analyses. The Sciex ICP-MS can be programmed to survey up to 16 different m/e ratios in a given scan, although the signal/noise ratio will obviously decrease as less of the time is spent at any particular m/e ratio. In Figure 2a the slugs of analyte are injected into a distilled water flow stream and the expected behavior is displayed. The 50 μL injections spread with turbulent mixing, diffusion, adsorption, etc. Whereas the injection slug is nominally of 3 s duration in the flow stream, the peaks in the detector trace approach five times that breadth. This mixing and dilution is also accompanied by a deterioration in detection limit compared to analyses of a flowing stream. Our figures indicate a loss of approximately an order of magnitude in detection limit compared to the manufacturer's limits for flow stream analyses (11).

Figure 2b reveals the more significant analytical difficulties which occur with FIA under some conditions. There the only change from Figure 2a is that the flow stream is now 0.1 M ammonium acetate instead of distilled water. The peaks for cadmium are very similar to those from the injections into a distilled water stream, however, those for copper now tail very badly. We believe that in the case of copper the dinuclear neutral complex $Cu_2(OAc)_4$ forms and that this is partially adsorbed onto the teflon tubing between the injector and the nebulizer. Regardless of the correctness of the explanation, it is clear that quantitative analysis of copper in this latter flow stream is not possible.

A second problem which arises in this type of flow injection analysis results from reaction between the

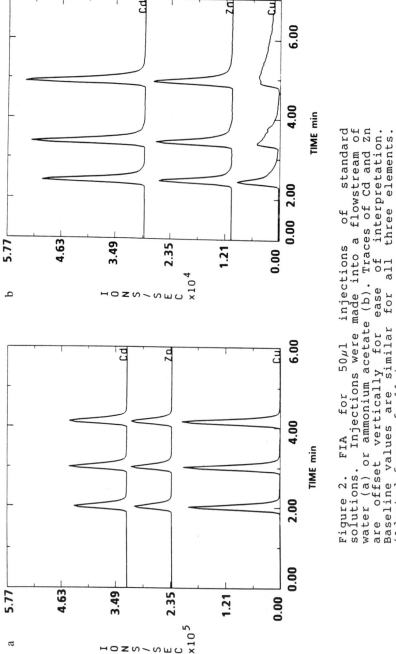

Figure 2. FIA for 50 µl injections of standard solutions. Injections were made into a flowstream of water (a) or ammonium acetate (b). Traces of Cd and Zn are offset vertically for ease of interpretation. Baseline values are similar for all three elements. (Adapted from ref. 11.)

analyte molecules and residual metal ions either in the injector or on the walls of the flow system. Thus, injections of blood plasma from a patient undergoing gold therapy show peaks for gold as expected. However, if a solution of bovine serum albumin, BSA, (which contains no gold) is injected immediately thereafter, gold is again detected. Either there is residual gold which is binding to the BSA, or the BSA is mobilizing residual human serum albumin, HSA, from the blood plasma which has gold bound to it. Again, the important point is that such effects can complicate any quantitative analysis and that the appropriate quality controls are especially necessary when dealing with trace levels of material in FIA (11).

Liquid Chromatography of Body Fluids

A wide variety of body fluids can be separated by liquid chromatography and characterized by various types of spectroscopy (12). These include urine, blood plasma, synovial fluid, cerebrospinal fluid, sweat, bile and amniotic fluid among others. Among these the most easily obtained are blood and urine and we have performed all of our experiments on these two types of fluid. For each of these fluids various types of chromatography are possible. For blood the normal procedure is first to separate either the serum or plasma and work separately with these. Note that for patients on auranofin therapy this removes nearly 50% of the circulating gold which is bound to the red blood cells (13). Blood plasma contains many proteins, the principal one being serum albumin. Most of the gold drug binding in vitro and in vivo is to the albumin. We have shown that in vitro this binding is to cysteine-34 of albumin (14). Presumably this is the in vivo binding site on human serum albumin as well.

Size exclusion chromatography of plasma or serum also allows separation of fractions containing the immunoglobulins. The immunoglobulins also carry some of the gold, albeit as a minor component. Figure 3 shows a typical chromatogram from a patient on auranofin therapy. In this case, element-specific detection has been used for Cu, Zn and Au and the UV detector trace has been superposed on the chromatogram as well. Note that the gold and zinc binding seem to conform extremely well, whereas the major peak for copper has a component which elutes before those of Zn and Au bound species. This pre-rise seems likely to include ceruloplasmin which has a molecular weight of ca. 120kD (15).

We have also used weak anion exchange (WAX) chromatography for serum and plasma. Currently we are exploring the use of internal surface reversed phase (ISRP) columns to separate unbound small molecules present in the plasma (serum) matrix. While transport of the gold-containing species seems likely to occur via the HSA, the active forms (or for that matter, the forms causing toxic side effects if different from the active forms) are likely to be small molecules. With the ICP-MS detector we have a FIA detection limit somewhat below 1.0 p.p.b. This corresponds to a detection limit for chromatography of below 10.0 p.p.b. Given gold levels in the blood of 1.0 to 0.1 p.p.m. for patients on gold therapy, this suggests that we can examine minor components present in blood but

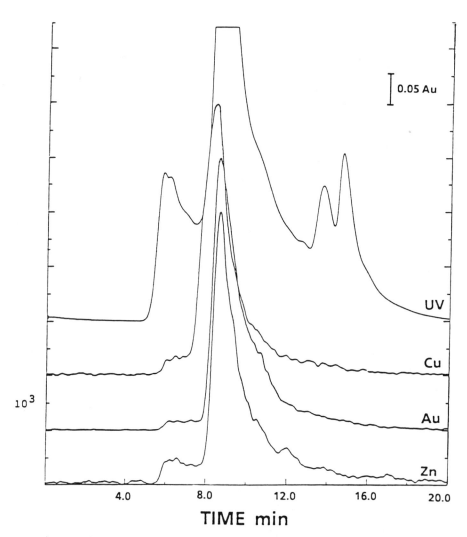

Figure 3. Size exclusion chromatography of blood plasma from an auranofin patient. The UV trace has been offset horizontally to account for the time delay in flow from the UV detector to the ICP–MS instrument. (Adapted from ref. 11.)

that they should constitute at least 1.0% of the material present. Below those amounts we will still need some step to concentrate the small molecules prior to separation.

HPLC of Urine Specimens

While blood samples provide one of the first possible metabolic specimens of the gold drugs, urine provides one of the final specimens in the metabolic chain. Since urine is also a readily available fluid, we have spent considerable effort in developing chromatography with the ICP-MS detector suitable for separating gold-containing metabolites in urine. Whereas the protein content of blood plasma or serum is high and most of the gold is protein bound, the amount of protein in urine is low, indeed clinicians monitor proteinuria as a symptom of nephrotoxicity from "heavy metal poisoning" associated with gold therapy (16). This leads to the expectation that most of the gold-containing materials in the urine are small molecules and suggests that the chromatography be adjusted accordingly.

We originally had our best success utilizing WAX chromatography on urine samples (17). Figure 4 shows the chromatographic results for separations on a series of samples obtained over time from a female psoriatic arthritis patient on auranofin therapy. (Psoriatic arthritis appears similar to RA except the rheumatoid factor is absent and skin rashes are present). The urine specimens were donated over a period of 21 weeks and frozen for subsequent analysis. The patient was initially taking 6 mg of auranofin daily. After 12 weeks the dose was raised to 9 mg daily and the patient's condition was stabilized.

Several comments concerning urine specimens are in order. First, the concentration of gold in the urine is expected to fluctuate with the fluid intake of the patient. In an attempt to compensate for this, we divide the gold concentration by creatinine concentration. Creatinine is a product of muscle catabolism and the amount produced is expected to be roughly constant for a given individual with time. Thus this normalization is an attempt to compensate for varying amount of fluid excreted. When we examined these samples for total gold to creatinine ratio, we found, as might be expected, a build up of the amount of excreted gold over time on therapy.

However the second point to be considered is, as has been noted by various authors, that there is no correlation between total gold content and drug efficacy (18). Our hypothesis is that given the variety of apparent metabolic pathways, efficacy and toxicity depend on the concentration of specific metabolites rather than total gold. Our own approach is twofold: to determine correlations between particular, uncharacterized metabolites and efficacy or toxicity; and also to separate and identify the various metabolites themselves. A third consideration with urine samples is whether samples are best chromatographed immediately on donation or stored frozen and all chromatographed at the same time from thawed

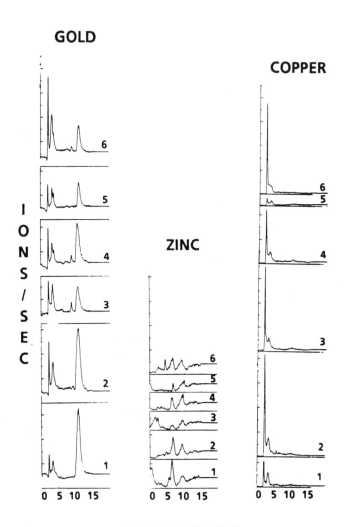

TIME (MINUTES)

Figure 4. WAX chromatography of sequential urine samples from an auranofin patient obtained over a 21 week period of treatment. For each sample indicated, a 50μl sample was chromatographed on a WAX column with simultaneous determination of Au, Zn and Cu. (Adapted from ref. 17.)

samples. We have chosen the latter approach, at least
while we are performing largely exploratory studies.

Returning to the results for urine chromatograms
shown in Figure 4, each of the samples was examined for
gold, zinc and copper. The six sequential samples have
been offset vertically with the earliest samples at the
bottom of the figure. The traces for gold, zinc and
copper are each presented in a different column for the
sake of clarity, although all three metals were scanned
simultaneously. The copper chromatograms indicate mainly
that WAX chromatography is not necessarily useful for
following copper distribution in urine specimens, since
most of the copper comes off the column in the void
volume, indicating no retention. One interesting
specimen is number 5. Very little Cu elutes from the
column (all vertical scales for a given element are the
same). Examination of the gold and zinc chromatograms
shows no such atypical behavior. Thus, the small amount
of copper is not the result of a bad column injection or
other such artifact. This aspect of internal quality
control is another advantage of scanning multiple metals
in a single chromatogram. Additionally there is a considerable time saving, and concerns about reproducibility
of temperature, mobile phase composition, pH, etc. over
multiple injections are avoided.

While the zinc results seem unexceptional, those for
gold show two interesting aspects. First, there are
multiple gold containing metabolites in the urine of this
single patient. Second, the metabolite distribution
appears to change with time. In the first chromatograms,
the major amount of gold is contained in the peak
retained for ca. 11 min. By the end of the series, the
unretained component and that at ca. 3 min are the
principal metabolites. Clearly the patient has "learned"
to metabolize the drug differently over time.

Use of Element-Specific Detection for Quantitative Gradient Chromatography

Many of the reported separations of biomedical fluids
utilize gradient reverse phase chromatography. This
causes considerable problems when using an ICP-MS
detector. First, the use of organic solvents is a
problem in and of itself. Nebulization of large amounts
of organic solvent into the plasma is likely to result in
extinguishing of the torch. Even if the power settings
are sufficient to keep the plasma intact, the formation
of elemental carbon is likely to clog the orifices to the
mass spectrometer with soot. Second, a changing
organic/aqueous composition, as is characteristic of
gradient C_{18} chromatography, exacerbates matters and
makes quantitative results more difficult to achieve.

We have adopted a number of modifications which
overcome these problems. Figure 5 depicts the changes we
have adopted to the nebulizer, spray chamber, torch chain
of our instrument. First, we have used a low flow
nebulizer which reduces the amount of material entering
the plasma. Second, we cool the walls of the spray
chamber to -20°C. to condense the organic vapors on the
walls and cause more of this material to be drained off
to waste. Third, we have a flow-metered oxygen inlet

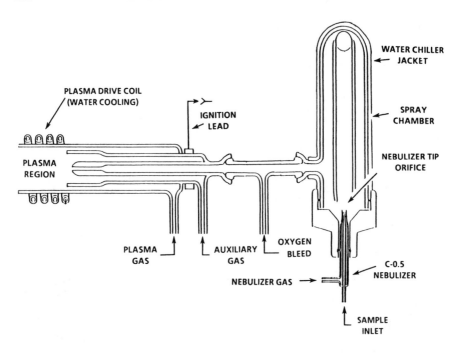

Figure 5. Modifications of the ICP-MS to facilitate use of organic solvents. The modifications include a low flow nebulizer, chilled spray chamber and oxygen introduction into the torch.

just before the plasma torch. This results in formation of volatile species such as CO and CO_2 and reduces the non-volatile soot buildup. Oxygen flow must be carefully adjusted however, as too much leads to rapid erosion of the ICP-MS orifices. We have found that aluminum sampler and skimmer cones can be machined locally and provide an economical source of replacements as needed.

The modifications cited above make it possible to run a gradient from 100% aqueous buffer to 100% methanol without problems of instrument failure. The ICP-MS detector results are not easily calibrated, however. Figure 6 shows the results of a flow injection experiment intended to simulate the results obtained from gradient chromatography. A gradient from 100% aqueous to 100% methanol was run over 10 min. Injections of constant composition slugs of a gold-containing drug were made approximately once per minute into this varying flow

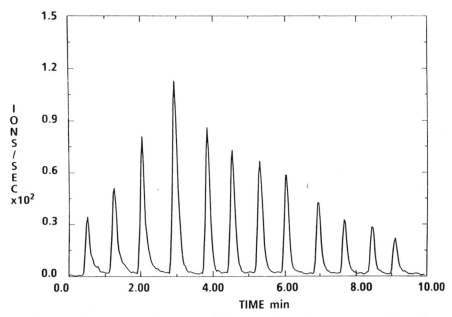

Figure 6. FIA of a gold sample into a methanol
gradient. 50μl injections of a 100 ppb solution of
auranofin were made into a flowstream consisting of a
gradient of 0-100% methanol.

stream.   As is evident from the figure, considerable
changes in sensitivity occur as the gradient progresses.
Thus, the ion count at a composition of ca. 35% methanol
is more than five times that for either 100% aqueous or
for 100% methanol.   Interestingly, this sensitivity curve
mirrors a plot of the relative viscosity of the mixed
solvent system as shown in Figure 7.   The results in
Figure 6 might suggest that optimal sensitivity could be
achieved by working with a 35%-65% solvent system.   The
occurrence of the maximum in the vicinity of 35% methanol
seems to occur independently of such instrument parameters as power setting of the rf generator or torch or
nebulizer flow rates, etc.   Aside from observing this its
generality.
      We have, however, provided a means, at small cost in
sensitivity, to avoid these problems.   Figure 8 shows a
modified HPLC system for input into the nebulizer of the
ICP-MS.   In it are added two extra HPLC pumps and a
second controller.   The analytical chromatographic column
is run in the normal fashion with each of the solvent
reservoirs, aqueous and organic, feeding one of the two
pumps and the combined stream of the desired composition,
determined by the controller, fed through the columns and
UV detector.   The second set of pumps and controller is
attached to the reservoirs in the opposite fashion and

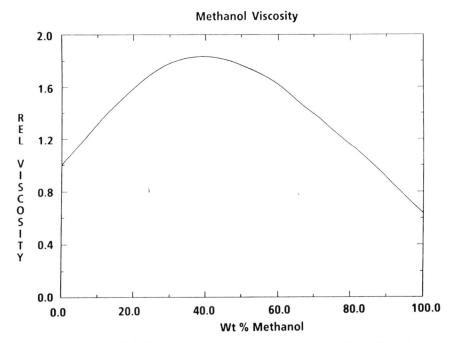

Figure 7. Relative viscosity of methonol/water solutions of different compositions. The vertical axis gives viscosity relative to pure water as a function of mobile phase composition. (Data from Handbook of Chemistry and Physics, Weast, R.C., Ed., CRC Press, Boca Raton, Florida, 66th Edition, 1985.).

the stream it produces has the reciprocal composition to the analytical stream. Thus when the column stream is 25/75 aqueous/organic the second stream is 25/75 organic/aqueous. These two streams at equal flow rates can then be joined in a zero dead volume tee to form the final analyte stream for the ICP-MS. While the cost is to reduce the amount of analyte in the final stream by a factor of 0.5, the result is to generate a constant composition stream of eluent matrix. Figure 9 shows the result of performing the same FIA experiment as in Figure 7 with this reversed composition, the second flow being added in. As might be expected, the peak to peak variation is slightly larger (ca. 4% rsd for peak height) than is achieved with a single stream of unvarying composition (ca. 3% rsd). This approach, while somewhat cumbersome and expensive, clearly provides an excellent solution for obtaining quantitative gradient chromatography.

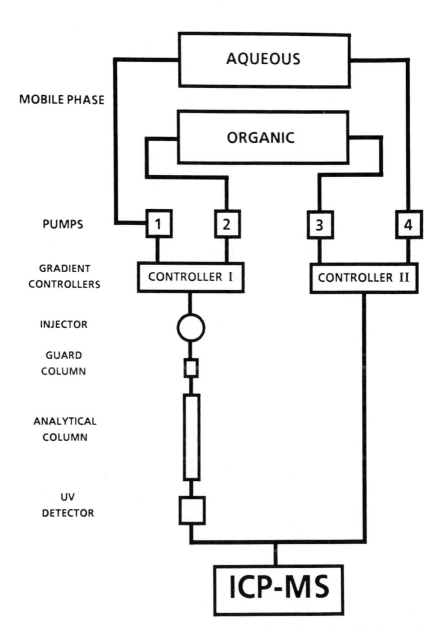

Figure 8. Schematic showing post-column addition of reversed gradient. This results in a constant composition flowstream entering the ICP-MS.

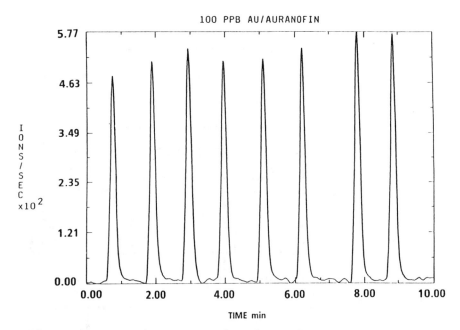

Figure 9. FIA into a methanol gradient using post-column reversed gradient. Auranofin samples were injected into a methanol gradient as in figure 6 but with the reversed gradient as described in figure 8.

Conclusion

ICP-MS provides currently the most sensitive element-specific detector for HPLC. Although expensive, it is extremely versatile in that it can be used in real time with the flow stream of an HPLC, it can monitor most of the elements of the periodic table, its sensitivity is two to three orders of magnitude better than ICP-AES, it can easily monitor several elements in the same injection and it can be adapted readily to monitor analytes in most of the currently favored solvents for chromatography. Considering the cost of other detectors, it would seem worthwhile for manufacturers to consider producing a "stripped-down" version of the ICP-MS specifically for this purpose.

Acknowledgments

The work we report here has been supported by NIH AR35370. The ICP-MS was acquired by the Biomedical Chemistry Research Center with support from a State of Ohio Academic Challenge grant. We thank Dr. S. Matz and M. L. Tarver for their contributions to this work. The idea of mixing a reciprocal composition flow stream into the eluent from a gradient separation was suggested by Dr. Alan R. Forster.

Literature Cited

1. Carter, S.K. In Platinum Coordination Complexes in Cancer Chemotherapy; Hacker, M.P.; Douple, B.; Krahoff, I.H., Eds.; Martinus Nijhoff: Boston, 1984; p 359.
2. Sutton, B.M. In Platinum, Gold and Other Metal Chemotherapeutic Agents; Lippard, S.J., Ed.; ACS Symposium Series 209; American Chemical Society: Washington, D.C., 1983; pp 355-369.
3. Blough, H.A.; Richetti,M.; Montagnier, L.; Buc. H. Second International Conference on Gold and Silver in Medicine, Manchester, England, April 1990, abstract p. 14.
4. Fox, C.F., Jr. Arch. Surg. 1968, 96, 184.
5. McCallum, R.I. Proc. R. Soc. Med. 1977, 70, 756.
6. Martin, J.L.; Yuan, J.; Lunte, C.E.; Elder, R.C.; Heineman, W.R.; Deutsch, E. Inorg. Chem. 1989, 28, 2899.
7. Holman, B.L.; Jones, A.G.; Lister-James, J.; Davison, A.; Abrams, M.J.; Kirshenbaum, J.M.; Tumeh, S.S.; English, R.J. J. Nucl. Med. 1984, 25: 1350.
8. Maxon, H.R.; Deutsch, E.A.; Thomas, S.R.; Libson, K.; Lukes, S.J.; Williams, C.C.; Ali, S. Radiology, 1988, 166, 501.
9. Elder, R.C. and Eidsness, M.K. Chem. Rev. 1987, 87, 1027.
10. Tepperman, K.; Finer, R.; Donovan, S.; Elder, R.C.; Doi, J.; Ratliff, D.; Ng, K. Science 1984, 225, 430.
11. Matz, S.G.; Elder, R.C.; Tepperman, K. J. Anal. At. Spectrom. 1989, 4, 767.
12. Bell, J.D.; Brown, J.C.C.; Sadler, P.J. NMR Biomed. 1989, 2, 246.
13. Sharma, R.P.; Smillie, J.; Palmer, D.G. Pharmacol. 1985, 30: 115.
14. Coffer, M.T.; Shaw, C.F., III; Eidsness, M.K.; Watkins, J.W., II; Elder, R.C. Inorg. Chem. 1986, 25: 333.
15. Takahashi, N.; Ortel, T.L.; Putnam, F.W. Proc. Nat. Acad. Sci. 1984, 81: 390.
16. Hess, E.V. Primer on Rheumatic Diseases, Ninth edition Schumacher, H.R., Jr. (ed.) Arthritis Foundation 1988, 93.
17. Elder, R.C.; Tepperman, K.; Tarver, M.L.; Matz, S.; Jones, W.B.; Hess, E.V. J. Liq. Chromatogr. 1990, 13, 1191.
18. Bluhm, G. J. Rheumatol. (Suppl. 8) 1982, 9: 10.

RECEIVED April 14, 1991

Chapter 19

A Fiber-Optic Spectrochemical-Emission Sensor as a Detector for Volatile Chlorinated Compounds

K. B. Olsen[1], J. W. Griffin[1], B. S. Matson[1], T. C. Kiefer[1], and C. J. Flynn[2]

[1]Pacific Northwest Laboratory, P.O. Box 999, Richland, WA 99352
[2]Department of Physics, Eastern Washington University, Cheney, WA 99004

A radio frequency induced helium plasma (RFIHP) detector was designed and tested as a sensor for volatile chlorinated compounds. The RFIHP detector uses a critical orifice air inlet and an RF-excited sub-atmospheric pressure helium plasma to excite the ambient air sample. The excitation source is coupled to a fiber-optic cable and associated collection optics to monitor the emission intensity of the 837.6-nm emission line of chlorine. The RFIHP detector demonstrated linearity from 0 to 500 ppmv carbon tetrachloride with a correlation coefficient of 0.996 and excellent reproducibility. The detection limit for carbon tetrachloride in air was 5 ppmv. Fluorinated compounds can also be readily analyzed by changing the analytical wavelength to 739.9 nm.

With the heightened concern about environmental issues that developed in the 1970s and 1980s, numerous laws were enacted to protect the human population and the environment from exposure to toxic and hazardous chemicals. Two of the more notable laws enacted by the United States Congress were the Comprehensive Environmental Response, Compensation and Liability Act (CERCLA), better known as "Superfund," and the Superfund Amendments and Reauthorization Act (SARA). These laws give the President of the United States authority to mandate cleanup of uncontrolled hazardous waste sites that are a threat to public health or the environment. As of November 1989, 1010 sites had been identified and classified as Superfund sites, with 209 additional sites proposed (1). Following identification and classification, the CERCLA process requires that a Remedial Investigation/Feasibility Study be conducted on all Superfund sites to determine the extent and types of contaminants present in air, water, and soil/sediment samples. Because of the sheer volume of samples sent to the laboratory for analysis, the cost to produce legal, defensible analytical data, the limited number of analytical laboratories available to conduct the analysis, and the SARA schedule deadlines, a reduction in the number of samples is

0097–6156/92/0479–0326$06.00/0

required. Clearly, the development of new screening methods for use by field teams is needed to decrease the number of samples being sent to the laboratory and to prioritize these samples.

The fiber-optic emission sensor offers many attractive features for real-time multipoint environmental field monitoring. These features include the small probe size, the multiplex advantage (i.e., multiple probes with one central detection and data acquisition system), and the potential for fast response. In this study, a sensor was developed that is capable of real-time, in situ monitoring for chlorinated hydrocarbon vapors in the vadose zone (region from ground surface to the water table) at Superfund and hazardous waste sites. A chlorine-specific sensor was selected because a study conducted by Plumb and Pitchford (2) concluded that a high percentage of hazardous waste sites across the country had volatile chlorinated organic compounds contaminating the ground water. Of the 15 compounds most often identified, 10 contain chlorine. This clearly suggests the prevalence of chlorinated hydrocarbon contamination at hazardous waste sites and indicates the need for a detector that could specifically measure chlorine-containing compounds.

Plasmas as atomic emission sources (AES) interfaced to gas chromatographs (GCs) and optical spectrometers have demonstrated their usefulness as element-specific detectors since the first demonstration by McCormack et al. (3) in 1965. Since then, the majority of the work cited in the literature on specific element detectors has been focused on the microwave-induced helium plasma (MIP) as the excitation source of choice for GC-AES (4-7). The GC-MIP system was found to provide sufficiently low detection limits and elemental specificity for most elements of interest. However, a major drawback of the MIP source was the plasma's inability to remain stable or lit when air was injected into the carrier gas stream. Operating a stable plasma with small but measurable quantities of air without destroying the excitation characteristic of the plasma is a critical requirement for a detector designed to measure total chlorine in vadose-zone air. In 1985, Rice et al. published a study using a low-frequency, high-voltage, electrodeless discharge-plasma (8). They found that this plasma had detection limits in the picogram range for a multitude of elements, including chlorine, phosphorus, sulfur, and mercury. They also found that the plasma was reasonably tolerant to the presence of contaminants and air. Additional advantages included the compactness of the plasma source and low helium consumption. Taking advantage of all the favorable characteristics of that system, we have designed a chlorine-specific detector capable of measuring volatile organic halogenated hydrocarbon compounds in air.

Experimental

A schematic diagram of the components used to construct the RFIHP probe is given in Figure 1. The operating parameters of this system are compared with those of the helium discharge-afterglow system reported by Rice et. al (8) (Table I). Major differences of the RFIHP system include using a 6-mm-o.d. x 0.5-mm-i.d. quartz tube as the plasma chamber, placing a 50-μm-i.d. x 10.2-cm-long segment of uncoated capillary column at one end of the plasma tube (used as a critical orifice), operating the plasma chamber at sub-ambient pressure, and viewing the plasma axially. The overall experimental system with specific components used is shown in Figure 2. A schematic of the optical

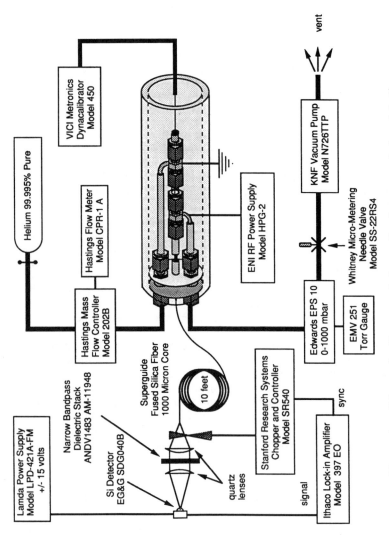

Figure 1. Schematic diagram of the radio frequency induced helium plasma detector experimental system. (Reproduced with permission from ref. 9. Copyright 1990 The Society of Photo-Optical Instrumentation Engineers.)

Table I. Comparison of Operating Parameters of
RFIHP Detector and Helium Discharge-
Afterglow System

Parameter	RFIHP	He Discharge-Afterglow
He Flow	40 ml/min	80 ml/min
Plasma Cell Pressure	100 Torr	Atmospheric
Critical Orifice Size	50 μm	N/A
Air Inlet Rate	0.68 ml/min	N/A
RF Power	23 watt Forward 18 watt Load	45 watt Incident 0.2 watts Reflected
Frequency	300 kHz	26-27 kHz

N/A = not applicable.

detection system appears in Figure 3. The absence of a spectrometer and the use of a narrow-band dielectric interference filter are important distinctions of this design. A transmission curve (verified in our laboratory) for this filter (Figure 4) is compared with the chlorine and oxygen spectral emission line over the wavelength range of 830 to 850 nm. Note that the narrow-band filter effectively isolates the 837.6-nm neutral chlorine emission line while rejecting adjacent emissions caused by atmospheric oxygen.

The silicon detector output is monitored by a synchronous (lock-in) amplifier (Figure 5). In this situation, significant stray RF emission arises from the plasma power supply; therefore, it is preferable to chop the fiber-optic output signal and utilize synchronous detection to eliminate false signals at the silicon detector output. The amplifier front end serves as a low-pass filter to effectively average the 300-kHz plasma optical output over a chopping cycle.

The probe concept was evaluated for use as a chlorine detector by monitoring the neutral chlorine emissions at 837.6 nm arising from the injection of chlorinated hydrocarbon or fluorochlorocarbon compounds into the probe. Detectability was determined for carbon tetrachloride, 1,1,1 trichloroethane (TCA), and dichlorodifluoromethane (Freon-12) (Figure 6). The RFIHP response to chlorine concentration was determined with a calibrated carbon tetrachloride vapor source consisting of a VICI Metronics (Santa Clara, CA) Model 450 Dynacalibrator equipped with one of the various sizes of diffusion vials available. Two different vials were sufficient to span the necessary concentration range (Vial #1: orifice diameter 0.5 cm, orifice length 3.81 cm; Vial #2: orifice diameter 0.2 cm, orifice length 7.62 cm). Two different diffusion oven temperatures were used: 46 and 66°C. Attainable concentration range was 0 to 1000 ppmv CCl_4.

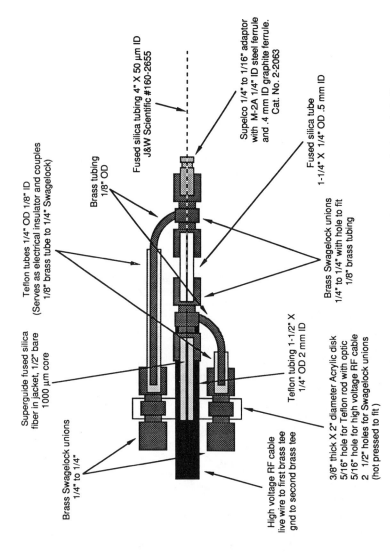

Figure 2. Schematic diagram of the radio frequency induced helium plasma probe. (Reproduced with permission from ref. 9. Copyright 1990 The Society of Photo-Optical Instrumentation Engineers.)

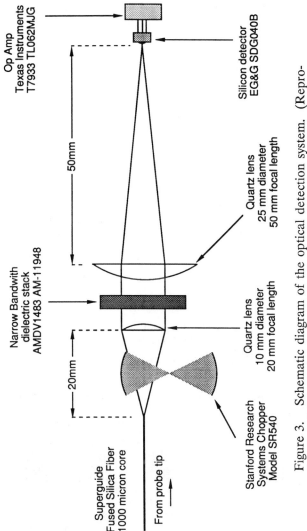

Op Amp
Texas Instruments
T7933 TL062MJG

Silicon detector
EG&G SDG040B

50mm

Quartz lens
25 mm diameter
50 mm focal length

Narrow Bandwith
dielectric stack
AMDV1483 AM-11948

20mm

Quartz lens
10 mm diameter
20 mm focal length

Superguide
Fused Silica Fiber
1000 micron core

From probe tip

Stanford Research
Systems Chopper
Model SR540

Figure 3. Schematic diagram of the optical detection system. (Reproduced with permission from ref. 9. Copyright 1990 The Society of Photo-Optical Instrumentation Engineers.)

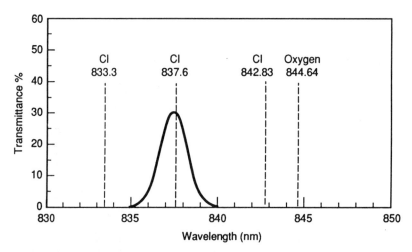

Figure 4. Overlay of narrow-band filter transmission curve and emission lines of oxygen and chlorine. (Reproduced with permission from ref. 9. Copyright 1990 The Society of Photo-Optical Instrumentation Engineers.)

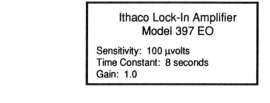

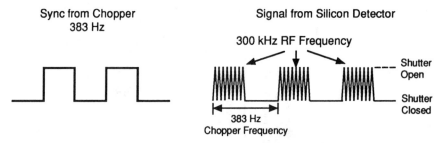

Figure 5. Electronic portion of RFIHP detector system. (Reproduced with permission from ref. 9. Copyright 1990 The Society of Photo-Optical Instrumentation Engineers.)

Results and Discussion

The RFIHP detector responds to any volatile compound containing chlorine, including dichlorodifluoromethane, carbon tetrachloride, and 1,1,1 trichloroethane (Figure 6). However, the relative response is dependent on the percentage of chlorine in the molecule. Therefore, the RFIHP detector is twice as responsive to carbon tetrachloride as to dichlorodifluoromethane. As with any detector based upon atomic excitation, this system would be sensitive to quenching caused by high concentrations of volatile organic species or excess air entering the plasma through the critical orifice. If excess air enters the plasma, the excitation potential of the helium plasma would degrade or be lost. However, no specific experiments have been conducted to study the effects of air concentration on plasma excitation properties. Quenching of the chlorine signal by self absorption of the emission line would be expected if high concentrations of chlorinated hydrocarbons entered the plasma. This would result in underestimating the actual concentration of the chlorinated species being measured by the RFIHP detector system. Another concern with a system based on atomic emission is from the effects of non-chlorine-containing compounds (hydrocarbons) on the chlorine signal (selectivity). To date, no specific experiments have been conducted to study these effects. However, we would expect a significant rise in continuum in this region of the spectra if percent levels of hydrocarbon compounds entered the plasma. If such an event occurred, a positive increase in the chlorine signal would occur, resulting in concentration estimates greater than the actual concentrations.

The RFIHP detector response over the range of 0 to 500 ppmv CCl_4 is plotted in Figure 7; the response is essentially linear over the entire range. Data scatter about the linear fit is attributed to uncertainty in the Dynacalibrator rotameter setting. Reproducibility of sensor output is demonstrated in Figure 8, where output for four sequential runs is plotted (Run #1 through Run #4). There is a downward trend in sensor output for zero concentration in sequential runs (the curves have approximately the same slope, ranging from 0.178 to 0.156); the cause of this trend is undetermined. This time-dependent, baseline drift is further illustrated for a different set of data in Figure 9, where Run #1 was initiated at high concentration and Run #2 was initiated at zero concentration (i.e., the zero concentration values for Runs #1 and #2 were taken sequentially). In this situation, the disparity in the curves at the high concentration values is consistent with baseline drift. The lower detection limit for CCl_4 can be estimated from Figure 9 as approximately 5 ppmv, which corresponds to a detection limit for chlorine of 4.6 ppmv in air. The RFIHP detector demonstrated a correlation coefficient of 0.996 over the range of 0 to 500 ppmv CCl_4 and good measurement reproducibility over the range of 0 to 200 ppmv CCl_4.

The primary use of the RFIHP detector at hazardous waste sites would be to measure the concentration of volatile chlorinated hydrocarbons in the vadose-zone air. A second application would involve indirectly estimating the concentration of volatile chlorinated hydrocarbons in the ground water using Henry's law constant. This estimate, however, assumes that the air concentration of chlorinated hydrocarbons directly above the water (within a few inches) is in equilibrium with the concentration in the water and that the specific species in the water is known. In Mackay and Shiu (*10*), the Henry's law

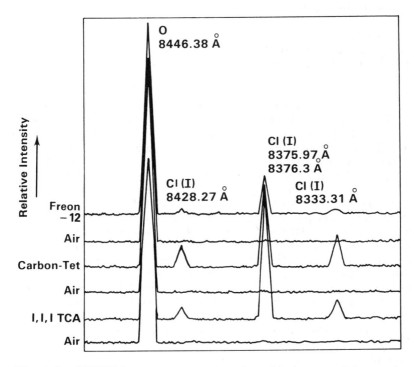

Figure 6.   RFIHP detector response to various chlorine-containing species in air. (Reproduced with permission from ref. 9. Copyright 1990 The Society of Photo-Optical Instrumentation Engineers.)

constant for carbon tetrachloride estimates a partitioning ratio between air and water at equilibrium to be approximately 0.95 to 1.0, resulting in an air concentration of 150 ppbv for each   g/L of carbon tetrachloride in the water. Therefore, the RFIHP detector could, theoretically, detect approximately 34   g/L of CCl_4 in the water. By applying a similar relationship to other volatile chlorinated hydrocarbons, the RFIHP detector could estimate their concentrations in ground water. However, it is uncertain whether perturbations introduced by the detector would cause deviations from Henry's law behavior.

Other industrially oriented applications of the RFIHP detector include monitoring for concentrations of selected species in air containing some of the following elements: mercury, phosphorus, or fluorine. The species could exist in the gas phase or as a dry aerosol. The monitoring could be related to hazardous vapor detectors and alarms or process control monitoring. For mercury and phosphorous, special consideration of the fiber-optic cable must be taken into account. Because the most intense emission wavelengths for these elements are in the blue or ultraviolet region of the spectrum, the fiber-optic cable must have the capability to carry that region of the light spectrum efficiently. Fiber-optic cables with this capability currently are very costly and still absorb a significant amount of light in that region of the spectrum.

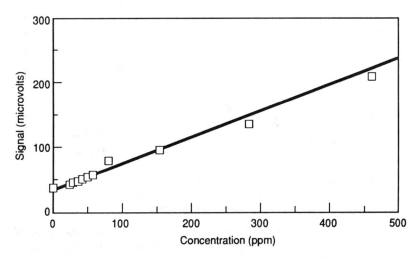

Figure 7.   Response curve for the RFIHP system over the range of 0 to 500 ppmv CCl_4 in air (837.6-nm emission line).   (Reproduced with permission from ref. 9.   Copyright 1990 The Society of Photo-Optical Instrumentation Engineers.)

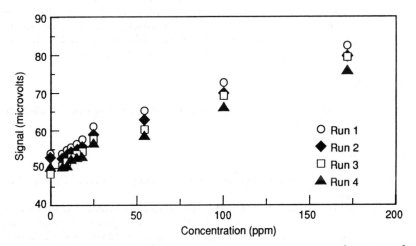

Figure 8.   Sequential RFIHP detector response curves over the range of 0 to 200 ppmv CCl_4 in air.   (Reproduced with permission from ref. 9. Copyright 1990 The Society of Photo-Optical Instrumentation Engineers.)

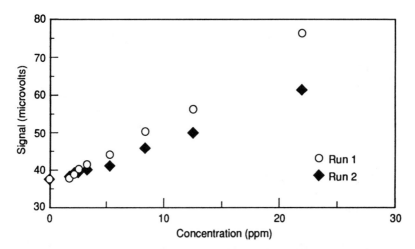

Figure 9. Sequential RFIHP detector response curves over the range of 0 to 22 ppmv CCl$_4$ in air. (Reproduced with permission from ref. 9. Copyright 1990 The Society of Photo-Optical Instrumentation Engineers.)

The RFIHP detector system has great potential for environmental monitoring applications where spot measurements are needed in the field or where real-time, in situ measurements are required for volatile chlorinated hydrocarbon compounds in soil gas or vadose-zone monitoring wells. Because no expendable materials are used by the probes, their lifetimes should greatly exceed those of fiber-optic probes based on colorimetric or fluorimetric chemical reactions. Consequently, the RFIHP probes are well suited for continuous environmental monitoring over extended periods of time.

Acknowledgment

This work was supported by the U.S. Department of Energy under Contract DE-AC06-76RLO 1830.

Literature Cited

(1) United States Environmental Protection Agency. National Priority List, November, 1989.

(2) Plumb, R. H.; Pitchford, A. M. *Proceedings of the Petroleum Hydrocarbons and Organic Chemicals in Ground Water - Prevention, Detection and Restoration - A Conference and Exposition*; November 13-15, 1985, Houston, TX. Sponsored by the American Petroleum Institute and National Water Well Association.

(3) McCormack, A. J.; Tong, S. C.; Cooke, W. D. *Anal. Chem.* **1965**, *37*, 1470-1476.

(4) Slatkavitz, K.; Uden, P.; Hoey, L.; Barnes, R. *J. of Chromatography* **1984**, *302*, 277-287.

(5) Slatkavitz, K.; Uden, P.; Hoey, L.; Barnes, R. *Anal. Chem.* **1985**, *57*, 1846-1853.
(6) Ester, S.; Uden, P.; Barnes, R. *Anal. Chem.* **1981**, *53*, 1829-1837.
(7) Quimby, B. D.; Sullivan, J. J. *Anal. Chem.* **1990**, *62*, 1027-1034.
(8) Rice, G. W.; D'Silva, A. P.; Fessel, V. A. *Spectrochimica Acta* **1985**, *40B*, 1573-1584.
(9) Griffin, J. W.; Matson, B. S.; Olsen, K. B.; Kiefer, T. C.; Flynn, C. J. In *Chemical, Biochemical, and Environmental Fiber Sensors*; Lieberman, R. A.; Wlodarczyk, M. T., Eds.; Proc. SPIE 1172; SPIE--The International Society for Optical Engineering: Bellingham, Washington, 1990, Vol. 1172; 99-107.
(10) Mackay, D. M.; Shiu, W. Y. *J. Phys. Chem. Ref. Data* **1981**, *10*.

RECEIVED April 26, 1991

Author Index

Affiliation Index

Subject Index

Production: Donna Lucas
Indexing: Deborah H. Steiner
Acquisition: Cheryl Shanks
Cover design: Amy Meyer Phifer

ʹuted and bound by Maple Press, York, PA